Can Bacteria Cause Cancer?

Can Bacteria Cause Cancer?

Alternative Medicine Confronts Big Science

David J. Hess

NEW YORK UNIVERSITY PRESS
New York and London

NEW YORK UNIVERSITY PRESS
New York and London

Library of Congress Cataloging-in-Publication Data
Hess, David J.
Can bacteria cause cancer? : alternative medicine confronts big science / David J. Hess.
p. cm.
Includes bibliographical references and index.
ISBN 0-8147-3561-4 (clothbound).—ISBN 0-8147-3562-2 (paperbound)
1. Carcinogenesis. 2. Cocarcinogenesis. 3. Bacterial diseases.
4. Cancer—Alternative treatment. I. Title.
RC268.5.H47 1997
616.99'4071—dc21 97-4910
CIP

New York University Press books are printed on acid-free paper, and their binding materials are chosen for strength and durability.

Manufactured in the United States of America

10 9 8 7 6 5 4 3 2 1

Contents

Acknowledgments

For comments, bibliographic references, and/or research help, I thank Jeanne Becker and the Memorial Sloan-Kettering Cancer Center, Gerald Domingue, William Fry, Patricia Huntley, Clinton Miller, Helen Coley Nauts and the Cancer Research Institute, Geronimo Rubio, Tom Rosenbaum and the Rockefeller Archives, and Jayme Treiger. I wish also to thank Gretchen Koerpel for her patience as a spouse and her reference help as a librarian. Steve Fuller provided helpful comments on the manuscript as a whole, and editor Eric Zinner provided very helpful advice on structuring and positioning the book. Given the controversial nature of the topics covered, it should be emphasized that the opinions expressed in this book are my own and do not necessarily reflect those of the people who have been kind enough to help me.

Caveat

The information presented here does not constitute medical advice for individuals; they are advised to seek the guidance of competent physicians.

1

Introduction

The carcinogen is the germ of our time. Most of the world's population lives in a sea of carcinogens: cigarette smoke, pollution, pesticides, asbestos, radiation, radon, excess sunlight, food additives, hazardous waste, poor nutrition, hormones, viruses . . . the list continues to grow. Because there is so much complexity and uncertainty regarding the risk factors and treatment, cancer is more than a disease. It is a political and social problem. Cancer is a symptom of a global civilization that is out of balance with its biology and ecology. It is a medical crisis that affects millions of individuals and their families, but it is also a political and scientific crisis.

Rates of cancer incidence in the United States have risen consistently at about 1 percent per year. In the late 1950s one in four persons was likely to contract cancer and one in five would die from the disease, but by the 1990s the incidence had climbed to over one in three with over one in four mortalities. Shockingly, current estimates indicate that in the United States 45 percent of males and 39 percent of females will be diagnosed with cancer in their lifetimes. Over one and a quarter million Americans are diagnosed with cancer each year, and over a half million die from the disease each year. It is possible that by the second or third decade of the twenty-first century half of all Americans will be diagnosed with cancer at some point during their lifetimes.[1]

Of the millions of Americans who have cancer at any one time, many have turned to alternative cancer therapies to complement their conventional treatments, to replace officially recommended therapies, or to use as a last resort when their doctors have given them up as terminal. When the hundreds of thousands of cancer patients who choose alternative therapies are recognized not as a single phenomenon but as a subgroup of the millions of other Americans who use alternative therapies for other diseases and conditions, the dimensions of a grassroots health care revolution become visible. This sea change is spurred by competition among health-maintenance organizations, which are driven by market forces to fund an increasing variety of alternative therapies. Likewise, constituent demands for better access to alternative therapies have led to legislative reforms in many states and at the federal level.[2]

This rapidly changing situation has given rise to a widespread need for evaluation: legislators, insurers, health-care professionals, and above all patients need more information about which alternative therapies work and which ones do not. Clearly, one way to answer this question is through massive public funding of randomized, controlled trials of specific alternative therapies that patients are now using. Constituent demands for this type of evaluation will grow as the use of alternative therapies grows. However, in addition to testing specific therapies, the broader theoretical frameworks also need to be evaluated. There is growing public dissatisfaction with cancer research and its annual expenditures of billions of taxpayer dollars. The emphasis on basic research and on toxic chemotherapy has led to few real success stories for conventional cancer treatment. Advances have centered on the early childhood cancers and some of the less common cancers; however, the advances associated with chemotherapy occurred decades ago and they probably cannot be extended to the more common cancers (Moss 1995). Overall five-year survival rates of about 50 percent increased only marginally from 1974 to 1987, and even those marginal gains may be due largely to earlier diagnosis.[3] Given this dismal picture for cancer treatment, advocates of alternative therapies often seek more than testing of specific therapies. They seek a reevaluation of current research programs and they demand funding for alternative ones.

This book contributes to the evaluation issue by providing a framework for examining not a specific therapy but a broader alternative research program in which therapies are embedded. The terms "research program" and "research tradition," which have been used in specific ways in the philosophy of science, will be used here loosely to refer to an interconnected network of empirical studies, research practices, applications (e.g., therapies), and guiding theories.[4] The fundamental question for this book is, "What is the best way to evaluate alternative medical research programs in order to improve the ones that are currently in place?" To develop an answer for this level of the evaluation problem, I focus on one alternative research tradition for cancer: work guided by the unorthodox theory that bacteria play an overlooked role in the etiology of cancer.

By assessing one alternative theory of cancer, I provide a general method for analyzing the politics and possibilities of alternative research programs. A key assumption is that in order for a criticism of existing research programs to be useful, it should assess the feasibility of an alternative. It does no good merely to criticize the failures of existing research programs; the more difficult and necessary task is to evaluate alternatives that may contribute to reconstructing the existing programs. For the complex field of cancer research, one possibility among many is research on the role of bacteria in tumor genesis and promotion.

Medical researchers have already changed their thinking on one major chronic disease, the gastric ulcer, which is now recognized to have a largely bacterial etiology. They have also shown an increasing interest in the role of bacteria in arthritis. Viruses are now widely recognized as the agents of a number of cancers, rare ones for humans and more common ones for animals. However, the standard explanations for reports of bacterial colonization of human cancer tissues are that bacteria represent either an artifact (contamination) or an opportunistic, secondary infection that plays little or no role in tumor genesis and/or progression. These explanations face a number of anomalous findings that will be described and evaluated. In effect, two theories will be compared: the theory that bacteria may play an unrecognized role in tumor genesis and promotion, and the null theory that bacteria found in cancer tissues are merely opportunistic, secondary infections that have little impact on the multistage process of carcinogenesis.

The comparison of theories should not be done naively. The much heralded "war on cancer"—which President Nixon declared when the Cold War was moving toward détente—has left behind a battlefield bloodied with the bodies of advocates of alternative therapies who have been victims of intellectual suppression. The history of research on the bacterial-etiology theory is no exception. In some cases the suppression may appear to be legitimate, because some of the research was so poorly described or executed that most people with some college-level science background would probably reject it as bad science. However, during the decades following World War II a large amount of legitimate scientific research has appeared in standard, peer-reviewed journals. These publications reveal some evidence that bacteria do play a role in tumor genesis or progression. Yet, the significance and importance of that role remain, in my opinion, unanswered. There are some reports of sera and vaccines based on bacterial cultures that show striking success. These therapies pose relatively low risk to the patient and relatively low cost for society to develop and produce them. Consequently, the bacteria and cancer research program appears to be one example of the many alternative cancer research programs that warrants legislative mandate from Congress, approval for testing by the

FDA, funding from public agencies, attention from cancer researchers, and evaluation by open-minded clinicians. Like other alternative cancer research programs, the bacterial program may turn out to be a dead end. However, additional evaluation can be obtained at relatively little cost to a citizenry that is already paying two billion dollars per year for cancer research and much more for toxic treatments.

More generally, the cancer research agenda needs to be rethought in light of the stalled progress in cancer treatment and the gap between research at the level of molecular biology and useful new therapies. The public needs to reevaluate existing cancer research funding with the prospect of diverting a substantial percentage toward projects of relatively low cost that demonstrate reasonable preliminary scientific evidence and a possibly high benefit to patients. Such a strategy has already been adapted, with some success, in the area of specific immunotherapies. In the conclusion, I will advocate extending the research and testing of immunotherapies to those based on the bacterial etiology theory. More generally, I will argue in favor of extended funding for evaluation of those alternative cancer theories and therapies that pass what I will describe as a credible biological mechanism test, provided that they can muster some clinical evidence for efficacy and safety.

Because anyone who dares to question the standard wisdom of cancer research may face severe and deliberate distortion through attacks by a well-funded network of debunkers, it is worth pausing for a moment to underscore the specific claims made in this book. I ask four basic questions in this book: what is the history of this field of research, what are the reasons for its rejection, how sound is the research, and what policy changes would make it easier for a more measured consideration of alternative research programs. My claims are very specific, and they correspond to the chapter organization of the book:

1. There is a history of research on bacteria as etiological agents in cancer, and this research tradition has not been merely forgotten or disregarded, but actively suppressed. This proposition is demonstrated in chapter two, which reviews the relevant historical record.
2. The explanation for the suppression needs to take into account factors beyond those of evidence. Chapter three demonstrates that financial and professional interests as well as more general cultural factors are necessary to explain the history of suppression.

3. When one reviews the evidence in support of the bacterial etiology theory against today's best available scientific knowledge, it appears that bacterial infections may play some contributing role in the genesis or promotion of some human cancers. Furthermore, the occasional efficacy of bacterial vaccines—a clinical finding that has long been recognized in cancer research—may be due partly to their effect on bacterial infections, not merely because they provide general stimulation to the immune system. However, the fourth chapter also considers counterevidence and counterarguments, and some of the earlier claims—for example, that there is a single "cancer microbe"—are rejected as probably erroneous.
4. Several other alternative cancer therapies show some preliminary evidence in support of claims of safety and efficacy, and they have a credible possible biological mechanism. Those therapies should be provided with additional public funding for evaluation. The fifth chapter argues that in the United States (and perhaps other countries), the funding, evaluation, and regulatory apparatus for cancer research needs to be reformed substantially so that reasonable alternative therapies are evaluated promptly and fairly.

It is likely that some readers will misunderstand the point of the book as an argument that bacteria cause cancer. The scientific-medical argument that I make is much narrower. In order for the question posed in the title of this book to make sense, it is necessary to view the role of bacteria as *contributing* agents to cancer causality, perhaps even as promoters rather than initiators. In other words, a more scientific way of asking the question would be: "In human bodies that are immunologically compromised by poor diet, exposure to carcinogens, and other risk factors, can the emergence of latent bacterial infections contribute to tumor genesis and/or promotion?" I argue that bacterial infections may play an overlooked role in the etiology of cancer, and understanding that role could be important for developing more effective therapies.

The case of the bacterial theory is merely one among several very promising approaches that warrant extensive public funding for further investigation as adjunctive treatments or in some cases replacements for conventional therapies. The other alternative therapies include some of the dietary, herbal, metabolic, and immunological approaches. It is very possible that the bacterial theory is not even the most promising of the many alternative research programs in cancer research. Rather, the case study of the bacterial theory provides a method for a holistic analysis—sociological/anthropological, biomedical, and policy-oriented—of theories and their affiliated research programs in the alternative medicine field.

This case study also provides an empirical example of an alternative type of analysis for those involved in the interdisciplinary field of science and technology studies—the philosophy, sociology, history, anthropology, cultural studies, and policy analysis of science and technology. For those who are interested in the science-studies theory that informs my analysis, the key arguments are developed in the appendix. I advocate moving beyond the assumptions of the neutral observer in science studies to the scientific evaluation of research programs as policy choices, and of public institutions as candidates for reform. Thus, the voices sounded here are not merely those of descriptive science and social science, but that of a citizen who is concerned with policy reforms of great public concern.

2

Germ Warfare

The Case for Bacteria as Carcinogen

If anyone were to claim today that there is a causal relationship between bacteria and cancer, most cancer researchers would quickly dismiss the idea. Such skepticism is the product of a history in which some researchers claimed that bacteria and/or viruses were the sole etiological agents of cancer. Against such unicausal theories for a disease—or variety of diseases—as complex as cancer, skepticism was warranted. However, the skepticism may have overcompensated for the more extreme claims. A review of the history of claims of bacteria as carcinogenic agents may lead to an intermediate position between extreme skepticism—there is no relationship between bacteria and cancer—and extreme advocacy—cancer is an infectious disease like tuberculosis, caused by a single bacterial species. The first step in this exercise is to demonstrate that some of the advocates of a bacterial theory of cancer had a less than fair hearing.

One of the classic expressions of extreme skepticism was the first edition of James Ewing's *Neoplastic Diseases* (1919). In that book the influential director of what was then known as the Memorial Hospital (now Memorial Sloan-Kettering Cancer Center) pronounced his diagnosis: "The parasitic theory . . . appealed to the ancients, was tacitly accepted throughout the Middle Ages, was definitely argued by modern observers, and reached the height of its popularity as a scientific theory about 1895, but during the last fifteen years it has rapidly lost ground, and today few competent observers consider it as a possible explanation of the unknown element in blastomatosis" (1919: 114). Ewing's position on microbes was an extreme one. He rejected a contemporary study of a bacterium that brought about tumorlike growths in plants *(Agrobacterium tumefaciens),* and he also rejected the work of Rockefeller Institute researcher Peyton Rous on chicken sarcoma viruses (117, 121). At most, he argued, studies of the microbial origin of cancer suggest that a microorganism "may have a special capacity to excite inflammatory processes which tend to go on to tumor growth, but they offer no support to the theory of a specific cancer parasite living in symbiosis with the cancer-cell and constantly stimulating its growth" (125).

Although Ewing reported that interest in the microbial theory of cancer had been dying out after the mid-1890s, viral oncologists today view the first decade of the twentieth century as the originary moment of their field. As Ewing's comments suggest, recognition for their field was slow. It grew gradually over the decades as studies mounted to show evidence for viral etiology in a number of cancers. Rous was not honored with the Nobel Prize until 1966; as a field, tumor virology did not become an important part of mainstream cancer research until the 1960s, when successes with the polio vaccine sparked a renewed interest in viruses and cancer. Nevertheless, although virology is now an accepted part of cancer research, viruses today remain relegated to the rarer human cancers, such as cervical and skin carcinomas, Kaposi's sarcoma, and some kinds of leukemias, lymphomas, and liver cancers.

Another of Ewing's mistakes involves *Agrobacterium tumefaciens.* The bacterium is now recognized as producing tumorlike growths in plants. Furthermore, microorganisms today are known to be involved in carcinogenesis through mechanisms other than inflammation, as in the case of the fungal carcinogen aflatoxin. However, these examples are regarded as relatively limited causal linkages between cancer and nonviral microorganisms.

Thus, in general terms Ewing's view of the importance of microorganisms to the understanding of cancer remains the consensus, especially for bacteria and fungi. The consensus today is that bacteria isolated from cancer tissues represent secondary infections that have little if any role in the promotion of tumor growth. Bacteria are opportunistic, not etiological, agents. Where they are recognized today as playing a role in the etiology of human cancer, such as the case of the ulcer bacterium *Helicobacter pylori,* it is assumed to be through chronic tissue irritation that may set the stage for tumor genesis. Likewise, although there is a tradition of immunotherapies that use bacteria and bacterial products to stimulate the immune system to attack cancer cells, the efficacy of these therapies is believed to rest on general immune system stimulation rather than a specific attack on purported oncobacteria. Leading tumor immunologists such as Steven Rosenberg therefore are suspicious of the apparently "blunt and empirical approach" of bacterial vaccines such as the tuberculosis vaccine BCG, which has sometimes been used in the treatment of cancer. In his words, "[T]he immune system is highly specific; a response to one stimulus, the BCG, should not stimulate a response to another stimulus, cancer" (Rosenberg and Barry 1992: 59). When used for cancer, the old bacterial vaccines such as BCG are considered nonspecific because they work by stimulating the immune system in general. In contrast, specific immunotherapies such as interleukin-2 are considered superior because they attack cancer cells through recognition of specific immunological markers. Unfortunately, the new, specific immunotherapies suffer from problems of toxicity and only moderate efficacy.

Bacterial vaccines have sometimes been associated with the theory that bacteria, and sometimes fungi, are etiological agents in cancer. This theory was supported by a rich alternative research tradition that involved at least fifty scientists and clinicians in a number of countries. Popular during the nineteenth century, the theory received continued support during the twentieth century as a minority tradition. Although the quality of the research is very uneven, some of the best of the research has been published in recognized, peer-reviewed scientific journals. This alternative research tradition presents several anomalies in the received wisdom on bacteria and cancer. By no means do the anomalies require adopting a unicausal bacterial theory of cancer, as some of the researchers wanted. However, a broadened cancer research program that addresses some of the anomalies and some of the experiments of the bacterial research program could provide new and crucial insights into the etiology and treatment of cancer.

This chapter will demonstrate that many of the researchers suffered instances of intellectual suppression, particularly when they developed clinical applications. To accomplish a complete analysis, it is necessary to separate the "is" and "ought" questions. First, the demonstration that there *was* a systematic pattern of suppression needs to be answered before the question of whether such suppression *should* have occurred. Some readers may willingly grant the argument that such theories were suppressed, but they may then argue, "So what? The theories were so crazy that any reasonable person with a scientific background would also have advocated their marginalization. Scientists and doctors were merely doing their job, which is to protect the public from quacks." To some extent I will agree with the argument, but certainly not in such a baldly stated and oversimplified form. Among the cases that I will discuss there is a range of credibility and scientific rigor, from Coley at one extreme to Lorenz and Rife at another. Some of the advocates of bacterial approaches to cancer invited their own demise with reckless claims, illegal practice of medicine, overly secretive behavior, and sheer arrogance. However, others such as Coley and Alexander-Jackson were careful researchers who did not get a fair hearing from their peers. The sorting out of the wheat from the chaff is a task that I will defer until the fourth chapter, after I have first demonstrated a pattern of suppression and then provided an explanation of why it occurred.

Coley

The choice to begin with the American surgeon and cancer researcher William B. Coley may be controversial because the connection he drew between bacteria and cancer was primarily therapeutic. However, Coley supported a microbial theory of the etiology of cancer, and his therapy is certainly consistent with the theory. In a report on a conference in 1926 in which bacterial theories were heavily criticized, Coley wrote, "To those of us who believe that cancer is of infectious nature and due to some microbic organism or group of closely allied microorganisms, the discovery of such cause or causes offers the only real hope of solving this greatest of medical problems" (1926: 225). In other publications as well as personal correspondence, Coley defended the microbial theory as a contender worthy of serious research attention (1925, 1928, 1931).

Another reason to begin with Coley is that he has become recognized as a founder and precursor of contemporary cancer immunotherapies. An example of Coley's new status in the history of cancer research is a chapter by the prominent cancer researchers Herbert Oettgen and Lloyd Old on the history of immunotherapy in *The Biologic Therapy of Cancer,* where Coley is granted the status of a founding figure (1991: 97). He was a Harvard-educated surgeon born in 1862 whose first patient was a nineteen-year-old woman with bone cancer. After he amputated her arm, a standard treatment for the day, she died of widespread metastases of the bone cancer (sarcoma). According to his daughter, Helen Coley Nauts, D.Sc., hon.,

> This experience saddened Coley and made him recognize that he did not know enough about cancer, and so he went to the record room of the New York Hospital where he had interned for two years, and studied all the cases of sarcoma treated in the preceding fifteen years. He found a patient who had recovered from an advanced case of sarcoma after developing erysipelas in the wound following a fourth incomplete surgical removal. Erysipelas is a form of *Streptococcus pyogenes* affecting the skin, causing high fever. Coley succeeded in tracing the patient in 1891 and found he was alive and well,

> having had no further recurrences, seven years after his recovery. In 1891, Coley began inoculating erysipelas cultures in cancer patients and soon found it wise to use killed cultures since it was difficult to actually produce erysipelas or if one did to control it. Strep alone was not effective so he added another organism, *Bacillus prodigiousus* (now known as *Serratia marcescens*). This mixture was sterilized either by heat or filtration and was called the Coley toxins. The first case treated in January 1893 was a nineteen-year-old boy with an inoperable sarcoma involving the abdominal wall and bladder, who was incontinent and bedridden. This large tumor regressed completely under four months of injections and the patient remained well until death from a heart attack seventeen years later. During 1891–92 Coley studied the literature (German and French) and found that a number of German physicians beginning with Fehleisen in 1882 had actually tried to inoculate erysipelas into inoperable cancer patients.[1]

The results of Coley's vaccines were mixed but promising, and they were possibly better than overall five-year survival rates today for similar cancers at comparable stages. His daughter Helen Coley Nauts subsequently abstracted and analyzed 894 toxin-treated cases and found that 45 percent of the inoperable cases survived five years, as did 51 percent of the operable cases (Nauts 1975, 1976, 1980).

Science studies analyst Ilana Löwy argues that there was "a sharp decline in the application of Coley's toxins from the 1910s on" (1993: 340). She argues that Coley himself attributed the decline to the amount of time and supervision that the method required. Nauts disputes the extent of the decline; she points to the large number of publications by Coley and other surgeons, the amount of recognition that Coley received during his lifetime (especially at his retirement in 1931), and the fact that many doctors and cancer researchers were experimenting with his toxins. As Nauts points out, the Coley toxins were being used by the Memorial Hospital and the Mayo Clinic, by Henri Matagne of Brussels, and by a number of other doctors in the United States. Patients were also being referred to Coley from France and England. Furthermore, his toxins were prepared in Europe by the Lister Institute in London from 1894 to 1943, and by the German pharmaceutical company Sudmedica from 1914 to December 1981. Nauts points out that the German preparation, Vaccineurin, was widely used in Europe, and its use included diseases such as neuralgia and arthritis as well as cancer.[2] Thus, the question is not why Coley's toxins fell into decline, but why they failed to take off and to become the prevailing therapy of choice for sarcomas and perhaps other cancers.

Löwy gives five main reasons, of which the first two were originally suggested by Coley and Nauts. According to Nauts:

> Coley himself believed that the method required considerable time and effort on the part of the oncologist. Nauts found that the results depended largely on the potency of the preparation used and how it was administered—i.e., the site, dosage, frequency, and especially the duration of treatment. The best results occurred when marked febrile reactions (102°–104° F) were elicited and treatment was continued for three to four months. Eighty-five percent of the osteogenic sarcoma cases receiving toxins for three to four months survived five to fifty years. The survival rate for surgery alone in that period was ten to twelve percent (Nauts 1975: ref. 315). What both father and daughter suggest is that the therapy was not difficult to use if the product was potent and correctly administered. However, clinicians found it difficult to get consistent results because they were not aware that they were often using weaker commercial products and couldn't achieve the correct febrile reactions.[3]

Nauts adds that the commercial versions of the Coley toxins manufactured by Parke, Davis, and Co. were weaker than those made by Coley's bacteriologists, and likewise the Lister Institute products were probably very weak and ineffective.[4] The lack of uniformity, together with the time and effort required, were deterrents to widespread use of the method.

According to Löwy a third factor that contributed to Coley's toxins' failure to become widely accepted was the empirical nature of the therapy. In other words, the therapy did not appeal to other medical researchers because there was no explanation of why it worked. This is a potentially important factor because it could have contributed to blocking Coley's ability to build a powerful network. Löwy notes that there was no breakthrough in the understanding of mechanisms until the 1930s and 1940s.[5] However, the evidence provided by Nauts that many clinicians seemed willing to try the therapy notwithstanding its empirical nature suggests that this reason was relatively unimportant.

Regarding the question of mechanisms, Nauts has subsequently outlined a number of means by which the therapy might work. Perhaps the most obvious mechanism was that the fever generated from the infection was high enough to kill cancer cells but not high enough to cause permanent damage to normal cells. The fact that cancer cells die at a temperature of 105–107°F, or slightly below normal cells, has been at the basis of hyperthermia therapy, which has gained increasing acceptance as an adjuvant therapy in recent years. Hyperthermia therapy was efficacious enough that in 1977 the American Cancer Society removed it from its Unproven Methods list, and in 1984 the Food and Drug Administration approved it (Walters 1993: 241). However, Coley recognized that hyperthermia could not be the only mechanism (1931: 615–16).

Nauts outlined a number of other specific mechanisms, including increased lymphocytes and the bacterial sequestering of iron or nutrients indispensable for tumor growth (1980: 15–19). Another line of research involves the triggering of cytokines, or chemicals that regulate cellular growth and function. In 1971 a team of researchers at Sloan-Kettering Institute led by Lloyd J. Old, currently medical director of the Cancer Research Institute and the Ludwig Institute for Cancer Research, discovered tumor necrosis factor or "TNF" (Old 1988: 60). Interest in TNF has played a major role in subsequent research on immunotherapies, and it has also been partly responsible for a revived interest in Coley's toxins during the 1990s. Old's research and that of immunologist Charles O. Starnes of Amgen suggest that Coley's toxins stimulate a cascade of cytokines—including TNF, interferon, and interleukins—that induce an immune response to cancer.[6] According to Nauts, if the cytokines are administered separately, they can be extremely toxic, in contrast with the Coley toxins, which are well tolerated.

Löwy provides two additional reasons, which she deems most important, for the decline of interest in the Coley toxins: the Memorial Hospital director, James Ewing, opposed the treatment, and in the 1920s radiotherapy displaced Coley's therapy as a treatment for inoperable cancers. Thus, when Ewing tolled the death knell of the microbial theory of cancer in his book *Neoplastic Diseases,* he was perhaps underplaying his own role as an active participant, rather than a mere chronicler, of the change underway. In one of the ironic twists of medical history, Coley was the person who brought radiotherapy to Memorial Hospital, as Nauts describes:

> Coley established the first X-ray machine in Memorial Hospital in 1901 paid for by two of his wealthier patients, as Memorial Hospital did not wish to spend money on such an unproven and unexplored approach. He was determined that every new idea deserved evaluation. Later on X-rays and radium helped to eclipse the Coley toxins, especially when Ewing became Medical Director at Memorial Hospital [in 1913]. Ewing then became very enthusiastic about radium and ruled that every ward case of bone sarcoma must receive radium prior to any other treatment. This continued for ten

> years and not one single case survived. Coley's private cases received the Coley toxins to prevent metastases after surgery. If the toxins were given for at least 3–4 months, 85% survived.[7]

Coley retired in 1931 but he did not pass into obscurity. He was honored with a huge banquet of over two hundred colleagues, friends, and family at the Waldorf Astoria Hotel. The speakers included his "old adversary James Ewing."[8] In 1935 Coley was named an honorary fellow of the Royal College of Surgeons in London, and obituaries written at his death in 1936 honored him as a pioneer and leader in cancer research as well as in other areas of medicine.[9]

Bradley L. Coley, who succeeded his father in 1931 as the head of the Bone Tumor Service at Memorial, and his associate Norman Higinbotham continued to use the Coley toxins for several types of bone sarcoma. As Nauts writes:

> B.L. Coley and Higinbotham served in the Army from 1942 to 1946. Upon their return chemotherapy soon became a priority. Cornelius Rhoads, Medical Director of Memorial Hospital, wrote Parke, Davis, and Company in 1950 that they no longer needed to prepare the Coley toxins as they had been made in Sloan Kettering since 1946. (Parke, Davis, and Company had prepared the toxins since 1899). For a time Rhoads had the toxins made at Sloan-Kettering Institute and eight reticulum cell sarcomas were successfully treated with this product, combined with X-ray therapy in some cases. The limb was saved in all these patients; one was a Mayo clinic case (Miller and Nicholson 1971). Without warning Coley, Higinbotham, the Mayo Clinic, or anyone else, in October 1995 Rhoads arbitrarily stopped the production of the toxins at Sloan Kettering while several patients were under treatment. Rhoads had become enthusiastic about chemotherapy and wanted to drop the toxins altogether. At that time several patients were receiving them and it was a cruel blow.[10]

Rhoads also had intervened in 1941, when he wrote to John D. Rockefeller, Jr., and urged him not to fund the research of Helen Coley Nauts on the Coley toxins.[11]

By 1940 Nauts had begun assembling and correlating all the histories of the toxin-treated cases by Coley and other surgeons in the United States and abroad. In doing so she analyzed the factors that affect success and failure: the preparation used; the site, dosage, frequency, and duration of injections; the amount of fever produced; and the stage of the disease when the toxins were begun. Between 1953 and 1984 she edited or wrote eighteen monographs, some of which were published in medical journals such as *Acta Medica Scandinavica* and some by the Cancer Research Institute (e.g., Nauts 1975, 1980). She also presented papers in the United States and in several foreign countries. With help from Oliver R. Grace she founded the New York Cancer Research Institute in 1953. The Institute published monographs on Coley's work, provided postdoctoral fellowships, and funded some experimental and clinical research.

Clinical studies during the 1950s and early 1960s demonstrated the efficacy of the Coley toxins.[12] However, in 1963 the FDA ruled that the Coley toxins had to pass through the complete new drug-approval procedure, and in 1965 the American Cancer Society put the vaccine on its list of unproven therapies. Although the ACS is a private, nongovernmental organization, its decision to list therapies as unproven has historically corresponded with lack of funding for research, official government disapproval of a therapy, and prosecution of doctors who use the unapproved therapy. Given the implications of the FDA and ACS rulings for future testing, it is not surprising that Nauts strengthened alliances with more orthodox cancer immunologists. She notes that she was very fortunate when in 1969 the head of immunology at the Memorial Hospital, Lloyd J. Old, became Medical Director of the New York Cancer Research Institute. In 1971 he reorganized the Scientific Advisory Council to include the world's leading cancer immunologists, and in 1973 the name of the institute was changed to the Cancer Research Institute. Nauts adds, "In 1975 at the urgent suggestion of the members of the Council and of Memorial Sloan-Kettering Cancer Center, the American Cancer Society removed the Coley toxins from the list of unproven cancer therapies."[13] Since that time the organization that she founded has been a major supporter of postdoctoral fellowships in cancer immunology (over five hundred in twenty-five years).

However, until recently there has been relatively little clinical research on the Coley toxins. According to Löwy, a single clinical trial took place at Memorial Sloan-Kettering Cancer Center in 1976 (Kempin et al. 1981, 1983). Löwy writes, "In this trial some improvement of survival and disease-free intervals [was] observed in lymphoma patients treated with MBV, but the long-term effects of this treatment were judged statistically insignificant, and the trial was abandoned" (1993: 345). Nauts explains the problem with the trial:

> They gave a single injection of MBV (Mixed Bacterial Vaccine, the name then used for Coley's toxins) combined with a considerable amount of

> chemotherapy. Initially the survival rate was better but at the end of three years the survival rates were no better for those getting toxins than those on chemotherapy alone. The single injection was totally inadequate. (No patient treated by Coley was ever given a single injection).[14]

More favorable evaluations of the Coley toxins appeared during the 1990s, when some clinical trials of the vaccine appeared in the peer-reviewed literature. As immunotherapies have grown in importance during the 1980s and 1990s, it appears that interest in the Coley toxins has also revived. However, Nauts noted that at the same time there has been the new development of black-market manufacture of the Coley toxins, that is, without FDA permission or the supervision of the Cancer Research Institute. These black-market versions of the Coley toxins are being used in the United States and other countries.[15]

The history of the Coley toxins represents a relatively mild case of intellectual suppression compared to the fates of some of the other advocates of alternative cancer therapies discussed in this chapter.[16] Coley was allowed to practice his therapy, but it also lost out to radiotherapy and later chemotherapy, both of which had more powerful supporters within the cancer research community. The FDA and ACS rulings appeared to seal the fate of the therapy in the 1960s, but the efforts of Helen Coley Nauts and her supporters, together with the increased interest in immunotherapies, have led to a comeback. Thus, the Coley case is interesting because it shows that a so-called unproven, unconventional, or alternative cancer treatment can be rescued. However, the price that has been paid is to classify it as an historical antecedent of specific immunotherapies, which are presumed to be more promising and more worthy of research attention today.

Glover

Coley's interest in the infectious theory of cancer led him to investigate the controversial claims of Canadian physician Thomas J. Glover. Glover is perhaps the most mysterious of the researchers who supported the infectious theory of cancer during the first half of the twentieth century. A highly secretive man, he became consumed with the belief that he had discovered the cure for cancer and that he would become rich and famous for the discovery. This belief reached such a level of obsession that he refused to share his knowledge with other researchers, and he cut off ties with anyone who might become a competitor for priority. Ultimately Glover did more harm than good to the infectious theory.

It is now known that the real research brain behind Glover was Tom Deaken, a laboratory assistant who had acquired a great deal of practical knowledge about science over the decades. Influenced by the ideas on cancer-causing microorganisms of Eugène Doyen of Paris and James Young of Edinburgh, Deaken began secret experiments on his own. Eventually he developed a medium upon which he was able to culture regularly what he believed was a cancer microbe. By 1910 he believed he could successfully inoculate laboratory animals and produce cancer with metastases in mice and dogs. Drawing on his research on a serum for hog cholera (research for which he did receive some credit), Deaken developed a serum for cancer as well as cancer vaccines, and by 1917–18 he had successfully used them on his experimental animals. (Of course, terms like "successfully" were more loosely defined in those days.) The serum was drawn from horses that had been inoculated with the supposed cancer organism (Boesch 1960: 234–41).

Notwithstanding his apparent success, Deaken could not interest doctors in his work, largely due to his lack of formal education and credentials. The first doctor who did pay attention to him was Glover. Glover and his business partner soon drew up an agreement with Deaken to form a corporation to market his products. Glover then returned to Toronto, where he had earned his medical degree, to open a cancer clinic that used the serum provided by Deaken. By October 1920 Glover was under investigation from Canada's deputy minister of health, and the department subsequently ordered him to make a complete presentation before the Toronto Academy of Medicine within three weeks. The committee appointed by the academy turned in a negative report in January, 1921, and Glover left Canada in order to pursue his research and fortune in New York (Boesch 1960: 240–44).

The Canadian committee claimed that Glover had been very secretive about his work. He refused to allow the members to visit his laboratory, to examine his cultures, or to witness any experimental demonstrations of his cultures or serum treatments (Boesch 1960: 205). It is likely that his reason for being so secretive was that he believed he had the cure for cancer and that he was going to get rich, and he did not want anyone to steal it from him. However, his strategy of secrecy produced opposition not only in Canada but also in the United States. In 1921 the *Journal of the American Medical Association* began publishing notices warning its members about the advertisements that Glover was circulating for his cancer serum. Francis Wood of Columbia University, a prominent cancer researcher, wrote that he purchased serum from Glover and injected it into white rats with transplanted carcinomas and sarcomas, and he concluded that there was no beneficial effect.[17]

Coley was among the few people who spent substantial effort to investigate Glover's claims fairly, but even he ultimately became very frustrated and disillusioned with the Canadian doctor. Coley first became aware of Glover and his serum in 1920, when he received a letter from a Canadian former patient.[18] In 1923 Archibald Douglas of the Memorial Hospital wrote to Coley about a report he had made on Glover, and Douglas offered to make the facilities available at the hospital if Coley and his colleagues were to conclude that there was anything of value.[19] However, Glover cancelled a visit from Coley and associates, and later Glover postponed the trial at Memorial indefinitely, claiming that he did not have enough serum.[20] Meanwhile, Charles Mayo of the Mayo Clinic wrote to Coley and suggested that a committee investigate Glover's claims. Mayo seemed to be genuinely interested in the research, and he wanted Edward Rosenow to look at it, but he was busy with other research.[21] Rosenow was an eminent bacteriologist who as early as 1914 had published on pleomorphism (bacterial form changes) within the streptococcus-pneumococcus group. Rosenow had long been interested in the theory that chronic diseases emerged from focal infections such as encysted wounds or infected root canals (Rosenow 1914; Hughes 1994).

Coley apparently continued to investigate Glover's claims by undertaking independent research at the Hospital for Ruptured and Crippled. Notwithstanding his attempts to win for Glover a fair hearing in the wider cancer research community, in 1931 Glover became concerned that Coley's research fellow, a physician named Richard Berg, believed that he had isolated a microorganism from malignant tissue that was different from Glover's organism.[22] Glover then wrote to Coley and claimed that he, not Berg, had isolated the organism from the malignant tissue, and that Berg had confirmed Glover's findings with a culture sent from his laboratory. Glover also claimed that Coley was calling the organisms the "Berg organism." Coley wrote back to explain that none of the charges was true and that he was in no way interested in taking credit from Glover.[23] Coley became increasingly frustrated with Glover's lack of cooperation, especially his failure to provide Coley and Berg with animals that had been inoculated with the organism that Berg and Coley had isolated from their chicken tumors.[24] Those animals were housed at Glover's laboratory due to the lack of facilities for extensive animal experimentation at the Hospital for Ruptured and Crippled.

Coley had many friends in England, including the prominent pathologist Sir Charles Ballance. He had persuaded Ballance to attempt to replicate the results that he and Berg had obtained, and Charles Mayo continued to express interest in the project. However, Glover's lack of cooperation put Coley in a difficult position. Finally, he wrote to Ballance and Mayo letters that explained the problems and frustrations he had experienced with Glover. His letter to Mayo stated that even if Glover were to provide him with a culture, Coley would not use it because he would not be sure that Glover had provided him with their own organism. As a result he decided to send to the British virologist William Gye for more of the "original dried virus," and to repeat the experiments by producing tumors in the fowl and attempting to culture the organism again. Coley added that Peyton Rous had urged him to dissociate his work from Glover's laboratory. Although Coley believed that all the years of working alone had made Glover especially suspicious, Coley nonetheless wrote that he still believed that Glover had found an organism of "great etiological significance."[25]

Apparently Coley persevered with his plan to start over. In 1935 he brought cultures to Gye, but a year later Gye wrote back a negative letter. He believed that the organisms were contaminants because he failed to produce cancer with them and because they grew easily on a simple agar medium.[26] However, as the British cancer researcher William Crofton noted, Gye was convinced of the virus-only version of the infectious theory, and his mind was closed to other interpretations (Crofton 1936: 108). In his letter to Coley, Gye reiterated his commitment to the infectious theory of cancer, but for viruses only. The available correspondence ends with this letter, and Coley died on April 16, 1936. Nauts wrote to me, "The awful anxiety and disappointments which Glover's behavior gave Coley over a ten-year period contributed to his duodenal ulcer, hemorrhages, and ultimately shortened his life."[27]

Another doctor who spent some time researching Glover's claims was Michael Scott, a surgeon from Butte, Montana. Scott became interested in Glover after he had used the new X-ray technology carelessly and developed cancer in his hand. His friend Rev. Charles Moulinier, S.J., the president of the Catholic Hospital Association, told Scott about Glover, so in October, 1921, Scott traveled to New York to meet with him. Glover told Scott that he had done all the research that in fact, as Scott later learned, Deaken had done. Scott became enthusiastic about Glover's research, and he made additional trips to New York. In the meantime, he had his affected finger removed surgically by a colleague in Portland, and he used the Glover serum as adjunctive treatment (Boesch 1960: 202–11).

On return visits, Scott followed Glover as they passed laboratory animals through the stages of Koch's postulates. Koch's postulates stipulate the conditions that are necessary and sufficient for establishing a causal relationship between a microbe and disease. Briefly, those postulates are as follows: the organism should be present in the diseased animals and absent from healthy ones; the organism should be grown in pure cultures; then it should be reinoculated into healthy animals and produce the disease symptoms; and finally upon re-isolation in pure culture the same organism should be found.

Glover remained secretive about the formula for the culture medium and the method for producing the serum. Scott wanted to make the research public, but Glover insisted on keeping it secret until he had tested the serum further. Scott continued to follow the research progress from a distance, and he sent Glover increasingly large sums of money to support his research. In the middle of 1922, Scott treated his first cancer patient with the serum. The woman had originally undergone a mastectomy for breast cancer, but the cancer returned and she was facing inoperable metastases. According to Scott, after six months of treatment with the serum, she was free of cancer. She lived until 1959, when she died from injuries due to a fall. Scott continued to treat patients, and by 1923 he had developed a network of doctors in the Midwest and on the West Coast who were testing the serum. In 1924 he took a leave from his medical practice and moved to New York. Scott also began working on publications of his results with the serum (Boesch 1960: 211–30).[28]

Scott continued to press Glover to publish his research results, but Glover refused. The tension led to a confrontation in 1926, when Glover became furious and accused Scott of "attempting to muscle in for a share of the riches that were sure to come" (Boesch 1960: 233). A few days after the split, Deaken visited Scott and told him the truth about his relationship to Glover.[29] Shortly after the falling out, Scott moved to Milwaukee, where he continued his research with the help of Father Moulinier, who tapped the wealthy Cudahy family to provide Scott with a laboratory and assistance. Deaken joined Scott in Milwaukee.

Father Moulinier then gave Scott top billing in the meeting of the Catholic Hospital Association in June 1926. Naively, Scott visited the AMA headquarters in Chicago and informed Morris Fishbein, the editor of the *Journal of the American Medical Association* from 1924 to 1949, of his forthcoming announcement. Scott made this courtesy call in order to give the AMA the first chance to make the public announcement about the cure for cancer. He must have been unaware of Fishbein's reputation. Today, he is a legendary figure in alternative medical circles; his epic battles with cancer herbalist Harry Hoxsey are well known, and he has been nicknamed in alternative medical circles the "medical Mussolini" (Beale 1939). Fishbein told Scott they were "too busy for anything like that," but he immediately began organizing a countercampaign. Eventually, Father Moulinier received orders from his superiors to withdraw Scott's paper from the program. Unable to disobey church orders, the Jesuit priest figured out a way around the orders: he closed the meeting early on the last day, held a short recess, then allowed Scott to speak unofficially. However, word was passed around the audience that Moulinier was out of his mind, Deaken and Scott were crackpots, and the whole business was a farce.

Shortly after the meeting, Father Moulinier was released from his position as head of the Catholic Hospital Association. Deaken and Scott returned to New York. Scott had become nearly penniless after having spent his fortune on cancer research. Moulinier and Deaken died within a few years, the latter believing that his ill health was from the cancer microbe that he had picked up in his laboratory work. Scott continued to travel when possible and to give lectures on cancer research at Catholic universities (Boesch 1960: 230–69). Before his death in 1967, a dentist in Butte, Robert Netterberg, became interested in Scott's research and managed to get a pharmaceutical company to attempt to isolate the microbes, but the efforts were without success (Netterberg and Taylor 1981: 9–11).

Glover and colleagues began publishing their results in the *Canada Lancet and Practitioner* in the mid 1920s, probably under some pressure from Scott and from Glover's business partner and funder J. J. Murdock. Glover claimed to have isolated a highly pleomorphic, or form-changing, microorganism that he believed was the etiological agent in cancer as indicated by his fulfillment of Koch's postulates (Glover 1926). He explicitly linked his research to related projects of some of his contemporaries by claiming that his organism was the same as that of cancer researchers James Young (1925a, 1925b), John Nuzum (1921, 1925), and C. Räth (1925). Glover also claimed that he could culture this pleomorphic organism from the Rous sarcoma virus, and thus that the filterable phase of his human cancer organism was the same as the avian virus.

Glover's work was distinguished from some of his colleagues in that he claimed to have isolated a complete microbial cycle for the cancer organism. Although microbiologists have long agreed that at least some species of bacteria can change form in response to environmental changes, the range of the possible changes and number of species that exhibit pleomorphism is controversial. Glover's position was located at an extreme end of the spectrum of opinion in two ways. He believed that the organism could pass through at least a dozen phases that included (in order) cocci, bacilli, spores, an amorphous substance (today sometimes called the L-phase), hyphae, spore sacs, and a filterable phase. The latter was a term that was used at the time for "viruses," which were distinguished from bacteria because they were able to pass through a filter that excluded bacteria. (At the time viruses were often called "filterable viruses," and their status as chemical agents or life forms was contested. Many also believed that viruses were merely one stage of a bacterial cycle.) In addition to the extreme pleomorphism that Glover attributed to the organism, he believed that the phases were organized in a life cycle, thus making him a "cyclogenist" (Glover 1930: 110).

Glover's 1926 publication in the *Canada Lancet and Practitioner,* which was heralded by a favorable editorial, presented a review of fifty patients. Most of them had achieved remarkable remissions, often from heavy tumor loads and for more than a year. It is clear from the report that a number of doctors throughout North America were either testing the serum or sending Glover patients, and he therefore had a fairly well-developed network of supporters that went well beyond the network assembled by Scott.

In 1929 Dr. George W. McCoy, the director of what was then called the Hygienic Laboratory of the United States Public Health Service, visited Glover's lab at the Murdock Foundation in New York City and invited him to repeat the work under supervision. The team included George Clark, a pathologist in Scranton who later became a key link between the Glover network and the post–World War II network of Virginia Livingston and Europeans such as Franz Gerlach. The work in what became the National Institute of Health began in December, 1929, and ended in 1938. In 1933 McCoy gave the go-ahead for the first report of the Glover team's research, which demonstrated the experimental production of a malignant adenoma with metastases in a guinea pig. As Glover recounts,

> Notwithstanding the fact that the announcement in the *Public Health Reports,* dated March 31, 1933 [Glover and Engle 1933], was rendered less harsh by

> the description of only one case, it had the effect of originating a wave of adverse criticism from what appeared to be an opposition organized against the acceptance of the microbic or virus doctrine of the etiology of malignant disease. This criticism, while aimed chiefly at the work, was also directed against those who sanctioned the work in the Institute and permitted the appearance of the above mentioned publication. These influences not only impeded the progress of the work in the National Institute of Health but deranged and delayed the publication of a manuscript that detailed the successful repetition, at the Institute, of the production of metastasizing tumors in other groups of experimental animals. (Glover and White 1940)

McCoy supported additional publication of their work, but there were delays. In 1937 McCoy was replaced by Dr. R. H. Thompson, who became director of the newly formed National Institute of Health. (The "s" was not added until 1948 when the National Heart Institute was created, according to Strickland 1972: 53.) Thompson appointed a special committee to review the planned publication, and after some delays the committee recommended publication of the material as three separate studies. At first the Surgeon General approved the recommendation, but then he added a member of the National Advisory Cancer Council to meet with the special committee and to review publication of the first paper again. After various negotiations and interferences, one of the minority members (presumably from the National Advisory Cancer Council) suggested the work should be repeated by a member of the staff of the Institute before publication, and the Surgeon General approved the suggestion (Glover and Engle 1938). Glover was extremely frustrated because from the beginning he had been supervised by McCoy, and McCoy had even said that he believed the quality of research was equivalent to that of his own laboratory. Because the new proposal of supervision would cause at least an additional two-year delay, Glover opted to return to the Murdock Foundation in New York and to publish his work independently.

For Glover the laboratory studies on guinea pigs were only an experimental portion of a total research picture that included his clinical work, which he regarded as more important. However, his request to the Surgeon General in March, 1938, to send a medical officer to his clinics and observe the results of the treatment was not acted upon. McCoy did spend several days at the clinics and told Glover that he was "greatly impressed, especially with those cases that were clinically free of the disease over a long period of time" (Glover and White 1940). Glover's 1940 report includes dozens of cases of long-term survival (over five years, sometimes over fifteen years) in patients diagnosed with cancer.

Glover's account of his experience in Washington suggests political intervention from cancer researchers who opposed the infectious theory. However, he helped to create the conditions of his own defeat. His secretive behavior meant that he lost key allies such as Scott and Coley, who could have helped shift opinion in cancer research circles. The Glover case is therefore complicated because he and Scott appear to have been victims of organized efforts to suppress their research, but Glover himself also suppressed recognition of the work of his assistant Tom Deaken and he was so concerned with priority that he lied to Coley about Berg's independent culturing of the organism. Glover's controversial research and clinical practice soon headed to the dustbin of medical history. By the 1960s the American Cancer Society's *Unproven Methods of Cancer Treatment* included a short description of negative aspects of Glover and his serum.

Rife and Kendall

Contemporary to Glover but on the West Coast, the Rife network involved an odd coalition of experienced medical researchers and an uncredentialed, independent inventor named Royal Raymond Rife. Like Glover, Rife claimed to be able to culture pleomorphic organisms from the tissue of cancer patients and to provide a nontoxic, successful cancer therapy. According to journalist Barry Lynes (1987), who worked with Rife's former partner John Crane to research the biography, Rife was born in Nebraska in 1888, served in the Navy during World War I, and was assigned to investigate foreign laboratories for the U.S. government. It is possible that he came across similar work by the researcher Georges Lakhovsky or that Rife met directly with Nikola Tesla.[30] After the war Rife worked as a handyman and chauffeur for the roller-bearing magnate Henry Timken, and he spent his productive adult research years in San Diego.

In his 1953 report Rife claimed that after 1920 he built several high-powered microscopes that worked at magnifications of 17,000X or higher, that is, at levels way beyond 2,000X to 3,000X associated with standard light and darkfield microscopes. He claimed to achieve this high level of magnification without sacrificing a great deal of resolution because he polarized light by passing it through rotating quartz prisms. Rather than stain samples, he used a variable monochromatic beam of light which he "tuned to coordinate with the chemical constituents of the particle, virus, or microorganism" (Rife and Crane 1953).

By 1931 Rife had attracted the interest of Arthur Kendall, the Director of Medical Research at Northwestern Medical School. Kendall brought with him his protein-based "K-medium" that allowed him and Rife to culture filter-passing bacterial organisms. By the end of 1931 Kendall and Rife published on the typhoid bacteria in the filterable state, which Kendall claimed they had been able to see with the Rife microscope (Kendall and Rife 1931; Kendall 1931). They also announced their findings before a meeting of several prominent doctors and researchers in Los Angeles, and Rife was soon demonstrating the microscope before the local medical and research community (Lynes 1987: 43–45). In 1932 Kendall presented his results before the meeting of the Association of American Physicians, and in that year he also published the articles in *Science* and the *Journal of the American Medical Association*. Kendall was an active participant in a scientific controversy that is sometimes called the "filtrationist" controversy; in the 1920s and 1930s bacteriologists debated claims that bacteria could pass through a filterable stage. Although Kendall failed to interest many of the leading bacteriologists in his research, one important exception was Edward Rosenow of the Mayo Clinic. In 1932 Rosenow met with Rife and Kendall at Northwestern, and he subsequently confirmed in print their observations of filter-passing organisms (Rosenow 1932; Lynes 1987: 46).

Kendall and Rosenow may have been more interested in Rife's research because of its implications for the theory that bacteria can enter into a viruslike "filterable" state, but Rife used his microscope to follow microorganisms through what he interpreted as their pleomorphic cycles. He soon arrived at an extreme version of the pleomorphic theory:

> We have classified the entire category of pathogenic bacteria into ten individual groups. Any organism within its group can be readily changed to any other organism within the ten groups depending upon the media with which it is fed and grown. For example, with a pure culture of bacillus coli, by altering the media as little as two parts per million by volume, we can change that microorganism in 36 hours to a *Bacillus typhosis* showing every known laboratory test even to the Widal retraction. Further controlled alternations of the media will end up with the virus of poliomyelitis or tuberculosis or cancer as desired, and then, if you please, alter the media again and change the microorganism back to a bacillus coli. (Rife and Crane 1953: 3)

By 1932 Rife believed he had identified viruses for cancer, typhoid, polio, and herpes. In that year he cultured bacteria from a breast mass that contained a ten-millimeter block of tumor that had been independently confirmed from another laboratory. After incubating the test tube for a day, he found it to be teeming with the cancer "virus." Rife repeated the method and claimed to get identical results. He also concluded from his experiments that thermal death was achieved at 107.6° F, the cancer microbe was sporogenous and anaerobic, and the virus was not destroyed by X-rays (at least the kind and dosage that he used). He believed instead that the dead tissue left by the X-rays formed a "natural parasitic feast" (Rife and Crane 1953: 11).

In accordance with Koch's postulates, Rife reinjected his viruses into rats and later found at the point of injection a mass that microscopic analysis revealed to be malignant. He repeated this procedure successfully over one hundred times (an older method sometimes used as a type of control), and he claimed that in other analyses he had isolated the cancer virus from a wide variety of human tumors. However, he did not think other laboratories would be able to replicate his procedure because conventional microscopes lacked the combination of high magnification and appropriate light frequency that were required to see the viruses. One solution to this problem was Rife's finding that by altering the medium in a slightly acidic direction, the virus transformed into something larger that would no longer pass through his filters. At the next stage, the organism reached a monococcoid form that he also found in the monocytes of the blood of over 90 percent of patients diagnosed with cancer. This form could be seen through a standard microscope when stained with silver nitrate and gentian violet. When cancer researcher and medical doctor O. Cameron Gruner from McGill University came to visit, they found that they could take the fungus that Gruner had isolated from his cancer samples, put it through the K-medium and filter, and arrive at Rife's virus; likewise, when they put Rife's virus on Gruner's pH-basic asparagus medium, it transformed into Gruner's "Cryptomyces" pleomorphic fungus (Lynes 1987: 71; Gruner 1942). This observation became the basis of a lasting friendship between the two researchers.

In the 1930s Rife also developed an electronic frequency instrument that could create the appropriate frequency to kill the virus (Rife and Crane 1953: 1). After successfully destroying the virus in over four hundred experimental animals by using the "mortal oscillatory rate" of his electronic-frequency machine, Rife began to experiment with human cancers.

He claimed that when he used the machine on a patient—today used by attaching electrocardiograph plugs to the patient—the machine would destroy or render harmless the organisms without causing any damage to the patient.

Under the supervision of medical doctor Milbank Johnson, a special medical research team was set up at the University of Southern California to test the Rife machine on humans. Johnson obtained funds from the Hooper Foundation for Medical Research at the University of California at San Francisco to carry out clinical trials (Lynes 1987: 55). They used the machine for three-minutes duration at three-day intervals in order to give the lymphatic system time to absorb and cast off the devitalized dead particles of the cancer virus. Rife and Crane write, "Sixteen cases were treated at the clinic for many types of malignancy. After three months, fourteen of these so-called hopeless cases were signed off as clinically cured by the staff of five medical doctors and Dr. Alvin G. Ford, M.D., pathologist for the group" (1953: 11). Unfortunately, there are no details on the definition for "clinically cured"; it probably means substantial tumor regression.

By 1937 Johnson had opened his third clinic, and there was widespread interest in the Rife "frequency instrument," especially in California (Lynes 1987: 71). Johnson and colleagues were finding that they could successfully treat a number of degenerative diseases, including cataracts. Rife also formed the company "Beam Ray" to begin manufacturing the instrument. However, Johnson and the special committee decided to hold back on making a public announcement regarding the efficacy of the frequency instrument, because they wished first to document the etiology of cancer, given the controversial nature of their claims (Lynes 1987: 83). Meanwhile Gruner, whose dean had denied him a leave to study with Rife, visited Washington, D.C., to examine the work of Glover, whose oncogenic microorganism Gruner believed to be the same as the Rife virus and Gruner fungus. Gruner wrote to Johnson that the Department of Public Health in Washington had undergone a change of management and it appeared that they would shut down the Glover research. He also wrote that he doubted that the distribution of the machine would make a big difference, because most cancer researchers did not examine living tissues and were not trained to culture microorganisms from cancer samples (Lynes 1987: 93–94).

The shutdown of the clinics and Rife research began when a patient they had treated returned to Chicago, and Morris Fishbein found out about the Rife machine. According to Lynes, Fishbein at first tried to buy in, and when that failed, he persuaded a disgruntled partner of Beam Ray to sue the company in 1939 (Lynes 1987: 89). Beam Ray eventually won the suit, and after the trial the judge offered to represent the defendants in a lawsuit against the AMA (97). However, the trial served its purpose. Rife ended up an alcoholic, his partners were left nearly bankrupt, and in the meantime any doctors who used the frequency emitters were threatened with loss of license. Immediately prior to the trial the "only other quality 'electronic medicine research lab' was mysteriously destroyed by fire" (99). The clinics were all closed down, and in 1942 Johnson sent his machine to Gruner, who decided not to use it out of fear of retribution. In 1944 Johnson died under mysterious conditions, which, according to Lynes, federal inspectors later ruled as death by poisoning (97).

Allies of Rife claimed that it was impossible to publish anything on the topic in medical journals. However, in 1944 Dr. Raymond Seidel became interested in the Rife machine, and he and Elizabeth Winter published a clever essay titled "The New Microscopes" in the *Journal of the Franklin Institute,* a scientific rather than a medical journal. The article took advantage of the interest in the new electron-microscopy technology to cover the Rife microscope along with the other technology. Seidel also described in an annual Smithsonian report how the frequency emitter could kill the cancer virus. Lynes writes, "Following the publication, Seidel soon became aware that he was being followed. Then a bullet crashed through his car windshield while he was driving" (1987: 98). Kendall was said to have been paid $200,000 to remain silent (102). Finally, a new technician stole the quartz prisms from Rife's microscope, rendering it impossible to operate. Rife closed the lab in 1947.[31]

After Rife became a recovering alcoholic, he entered into partnership with engineer John Crane to begin manufacturing the frequency emitters again. By 1960 they had leased out ninety machines to doctors across the country for testing. In that year Crane's office was raided, and equipment and records were confiscated. Crane was placed on trial for which the foreman of the jury was an AMA doctor, and he was denied access to his own confiscated records. Crane was found guilty and sentenced to ten years in jail, and he served a term of three years. Other doctors were told to stop using the machine, and Rife went into hiding in Mexico.

Rife died in 1971. The FDA continues the ban on the Rife machine for medical use. Rife generators and copycats are apparently available for those plugged into the appropriate networks; however, even within the alternative cancer-therapy movement, some have cautioned against the use of Rife machines because they are untested and they may be poorly calibrated. It is possible that the improper use of some bioelectric therapies may lead to cancer-cell proliferation rather than remission.

A Women's Network

Following World War II new networks of researchers and clinicians defended theories related to bacteria and cancer. One group of researchers centered on the controversial Krebiozen therapy, which like the Glover serum was developed by injecting a microorganism into horses. However, the microbe was *Actinomyces bovis* and the research was not positioned as a contribution to the study of a specific, pleomorphic, cancer-causing organism.[32] Another network, which obtained much more scientific credibility and which pursued the infectious theory, centered on Virginia "Livingston." Her last name is in quotation marks because it was as pleomorphic as the bacteria she studied: she is known in the literature as Wuerthele, Wuerthele-Caspé, and Livingston-Wheeler. Wuerthele is the name of her father, who was a member of the American College of Physicians. Her other last names reflect her ability to outlive more than one husband and her decision to change her name with each marriage. Although the changing names make it difficult to follow her work in the medical literature, she became known as a leader in the continuation and development of research on nonviral microbes in the etiology of cancer.

Livingston graduated from Bellevue Medical College of New York University in 1936 as one of four women doctors in her class.[33] Soon after graduating she met the commissioner of hospitals and complained that a woman had never been appointed as a resident or chief intern at a New York hospital. Ten days later she received an appointment as the first woman resident in New York City, where she worked in the infectious-disease section of a hospital prison ward for venereally infected prostitutes. Although the job was not what she had had in mind, she accepted it, "thinking that I would at least clear the way for future women residents." She adds, "My preconceived notions of the prostitute underwent rapid reevaluation, and I developed great compassion for these women, often diseased and discarded by society" (Livingston 1984: 56).

During World War II she worked as an industrial physician at Western Electric, and when she and her husband adopted a baby, she took a job as a school doctor in Newark. There, a school nurse with a skin disease asked Livingston to examine her. Although the nurse's doctor had diagnosed Renaud's syndrome, Livingston found signs of scleroderma, a progressive hardening of the skin that involves body organs and can be fatal. She had seen cases of leprosy and tuberculosis on her rounds as a resident in New York, and she decided to stain tissue samples from the nurse with the Ziehl-Neelsen stain, which was used to identify the bacteria of leprosy and tuberculosis.

Leprosy and tuberculosis bacteria are known as mycobacteria because they exhibit some characteristics similar to fungi. They are also known as acid-fast because they stain red and do not lose their color after being washed with alcohol. The method was discovered by Robert Koch in 1882 and was used to identify the tubercle bacillus, and acid-fastness is used today as a classification criterion for the mycobacteria (Atlas 1988: 284). When Livingston analyzed the stained samples, she found "an acid-fast microorganism that was neither the lepra bacillus nor the tubercle bacillus" (1984: 7). She reasoned that it was a new sclerobacillus and treated the nurse with medications for leprosy. The nurse, and later other patients with scleroderma, improved. Her curiosity aroused, Livingston set up a laboratory in her basement and studied the microorganism, which she found was pleomorphic but neither the lepra nor the tubercle bacillus. She published her work on scleroderma, and similar, confirming observations were made subsequently both in Europe at the Brussels Pasteur Institute and in the United States (Livingston, Brodkin, and Mermod 1947; Cantwell and Kelso 1971).

Livingston then found that when she inoculated animals with this microbe, many developed cancer. She began seeking tumor samples from colleagues in the area, and she found microbes that appeared to be the same as the scleroderma microbes. Concerned that she might have contaminated samples, she obtained tissue samples and blood directly from operating rooms and continued to find the same microbe in her slides. When the microbes were cultured and injected into mice, many developed cancer or a collagen disease such as scleroderma or lupus erythematosis (Livingston and Allen 1948).

During these years Livingston began a partnership with Dr. Eleanor Alexander-Jackson, who was working on the question of variation in the tubercle bacillus at the laboratory of Dr. Wilson Smillie of Cornell University. When Smillie found out that they were working on a possible bacterial origin of scleroderma and other collagen diseases, he thought of kicking them both out:

> However, one of his physicians challenged us with forty blood samples, of which some had been taken from patients with collagen diseases. We soon brought him a 100% accurate selection of the twenty-two samples that were infected. (Livingston 1984: 58)

After Livingston and Alexander-Jackson met the challenge successfully, Dr. Smillie's opposition softened. However, when Livingston began to suggest that cancer had the same etiology, Alexander-Jackson became concerned about the future of her work on the tubercle and lepra bacillus at the lab, and Livingston decided to stay away.

The president of the Newark Presbyterian Hospital then offered Livingston space for a laboratory in an old nurses' residence, provided that she had a university affiliation. Although there were no medical schools in New Jersey due to the state's antivivisection laws, she received an affiliation with the Bureau of Biological Research of Rutgers University. She began setting up the laboratory and received help from a number of foundations and laboratories, including the American Cancer Society and some pharmaceutical companies. She then assembled a team that included Alexander-Jackson as her first appointment.

One of their first activities was to obtain animal tumor samples infected with the major tumor viruses, such as the Rous, Walker, Sprage-Dawley, Shope, and Sarcoma-180 (Livingston 1984: 61). Believing that at least some viruses were filterable stages of bacterial cycles, she and her colleagues attempted to develop bacterial cultures from the viral samples and believed that they could do so successfully. In accordance with Koch's postulates, Livingston and colleagues produced diseases in animals from their bacterial cultures taken from virus samples. They used guinea pigs, which she stated produced cancer spontaneously in only one of a half million animals, but which produced cancer in a quarter of her samples. They also used genetically inbred mice that were known to produce a predictable percentage of tumors. The diseases ranged from tumors to tuberculosislike lesions to collagen diseases. Tissue samples revealed small cocci inside cells, which they believed were the intracellular cancer pathogen. In the course of these experiments they found that the disease could pass from animal drinking water that was contaminated with feces from diseased animals. Livingston and colleagues then reisolated the bacteria from the diseased animals (Livingston et al. 1950).

Livingston believed that their research proved that the Rous sarcoma virus and perhaps other cancer viruses were part of the cycle of an as-yet-undiscovered microbe. She later classified the pleomorphic, cancer-causing microorganism as part of the actinomycetales order of bacteria, and she christened the organism *Progenitoraceae cryptocides.* The name "Progenitor" reflects her belief that the organism is primitive and archaic, and the term "cryptocides" indicates its capacity to be a "hidden killer."

They argued that supporting evidence came from the research of Francisco Duran-Reynals. He came from the Pasteur Institute in 1926 to work on oncoviruses at the Rockefeller Institute, and after spending some time at Yale University went on to organize a Rockefeller-like institute in Spain (Corner 1964: 223–24). In the 1940s Duran-Reynals worked on the Rous virus and showed that it could be transmitted to a number of other avian species. In a review essay published in 1950, he argued in favor of a more extended view of viral variation to explain the various patterns of viral transmission. He believed that a number of non-neoplastic ordinary viruses could cause tissue destruction similar to or equivalent to neoplasms. In general, he thought that a key variable in the pattern of disease variation was the age of the host: "Infection of the old and generally more resistant host may manifest itself preferentially by cell proliferation rather than cell destruction" (Duran-Reynals 1950).

Livingston extended this position by arguing that the Rous virus did not behave like other viruses:

> A true virus has been defined as a submicroscopic infectious unit that lives only in the presence of living cells and cannot exist even momentarily outside of them. But Rous's "tumor agents" could be dried, stored on a shelf at room temperature for years, and when mixed with saline could then be reactivated to initiate fresh tumors. (1984: 65; see also Alexander-Jackson 1966)

In the 1950s she visited Peyton Rous and was warmly received. When she told him about growing the Rous virus in artificial media outside the living cell, "he said that he did not think this was unlikely or impossible" (Livingston 1984: 79). Rous had also argued that 95 percent of the chickens in New York City were infected, and he made the transmissibility of chicken cancer the subject of his Nobel Prize lecture in 1966 (Livingston

1984: 115). Thus, he seemed to believe that the chicken virus could cross species into humans. Livingston carried on his campaign by cautioning patients and readers about the possibilities of getting cancer from eating undercooked chicken, as in stir-fried chicken dishes.

Slowly Livingston's network of allies and fellow researchers grew. In about 1950 she got in touch with Elise L'Esperance, a doctor whom Livingston had first heard about in medical school when a professor disparagingly referred to a woman pathologist at Cornell University who believed Hodgkin's disease was caused by avian tuberculosis bacteria. In 1931 L'Esperance had published her report on Hodgkin's disease, in which she argued that Hodgkin's samples contained large multinucleated Reed-Sternberg cells that were similar to the giant cells of tuberculosis. A niece of the president of the New York Central Railroad, L'Esperance had been active in the New York Women's Infirmary for years. When her mother died of cancer, L'Esperance used the family money to found the Kate Depew Strong Memorial Cancer Detection Clinic in her memory. The clinic was the first of its kind to be founded in the world.

When Livingston visited L'Esperance's clinic, she met George Papanicolaou, the inventor of the Pap smear. As she notes, "Until L'Esperance demonstrated the usefulness of the Pap smear at her cancer detection clinics, Dr. Papanicolaou's work was not accepted" (1972: 51). When Livingston told L'Esperance about her work with acid-fast organisms similar to the ones L'Esperance had found, the latter said she had also isolated the organisms from the glands of the Hodgkin's patients, cultured them, and reproduced the lesions in guinea pigs after injecting the animals with the cultured organisms. She then showed the animal tissues to James Ewing, and he confirmed that they were Hodgkin's disease. However, "When she told him that they were from experimentally inoculated guinea pigs, he said that of course they were not Hodgkin's disease, that it was impossible to reproduce the disease from cultures" (1972: 52). Disgusted, L'Esperance nevertheless continued her work until a technician became ill and could not be replaced. She then switched to work in prevention and early detection.

By 1950 Livingston had also found support from Irene Diller, the editor of *Growth* and a cytologist at the Institute for Cancer Research in Philadelphia, and from her husband William Diller, a professor of parasitology at the University of Pennsylvania. Diller attempted to organize a symposium at the New York Academy of Sciences, but this meeting was blocked by the Memorial Hospital Director Cornelius Rhoads. He accused Diller of commercializing her work and therefore of not being qualified to sponsor a symposium (an ironic claim given Rhoads's relationship with the chemotherapy industry). All she had done was accept several ultraviolet sterilizing lights, with no strings attached, from a private company (Livingston 1984: 73–74). This event marked the beginning of the suppression that the Livingston network would face a lifetime battling. In November, 1951, the New York Academy of Sciences hosted a conference on viruses as causative agents in cancer that Rhoads organized (Miner 1952). The conference did not consider any of the research on bacteria and cancer, and Rhoads's introductory comments demonstrated that he was also skeptical of viral research as well.

Blocking the planned New York Academy of Sciences symposium was only the first of the run-ins between Livingston's network and Rhoads. When Alexander-Jackson was diagnosed with breast cancer in 1951, she underwent a radical mastectomy at Memorial Sloan-Kettering. While they were waiting, Rhoads called Livingston into his office and asked if they could try out a new surgical technique that involved splitting the sternum to remove glands around the heart and great vessels. "She would be performing a great service in permitting us to do this," Livingston states that Rhoads said to her, "as it would be an experiment to see how it would affect a patient and to determine the length of time she might survive" (1984: 75). Livingston was shocked and told him, "Not on your life! That is a cruel and disfiguring operation" (75). When Alexander-Jackson came through the surgery and Livingston told her what had happened, she was indignant. She resolved not to submit to any cobalt treatments or to any more surgery, and they resolved to use diet and the vaccines they were developing.

In 1953 Livingston and colleagues exhibited their work at the New York American Medical Association. The exhibit apparently created quite a sensation because a television hookup allowed visitors to watch the purported cancer microbes. "The publicity would have been great," Livingston writes, "but again the formidable Dr. Rhoads forbade the New York AMA publicity people to interview us. He also threatened to withhold further news releases from the press if they reported on our findings" (1984: 79). She claims that although there were crowds waiting to get into their booth, the press was intimidated and did not mention the exhibit at all in their reports.

Stymied in the United States, Livingston began preparations for the Sixth International Congress of Microbiology in Rome. Prior to the meeting she met with George Clark, the pathologist from Scranton who had worked on Glover's organism and had also produced tumors by injecting them into animals. Clark explained to Livingston that Glover had been able to produce antibodies in sheep and horses that were beneficial to humans with cancer. They tried to replicate the experiment but ended up infecting the sheep rather than vaccinating them. However, they tried again with a chicken farmer who was losing about a quarter of his chickens to fowl leukosis, and this time they used dead cultures to produce antibodies in rabbits. They experimented with six dying chickens. The two unvaccinated ones died and the four that received the vaccine returned to health (Livingston 1972: 43).

At about this time they raised funds to bring Dr. Franz Gerlach, a highly respected microbiologist from the University of Vienna, for a celebration in his honor. Gerlach worked on cancer and mycoplasma, a type of bacteria that lack a cell wall and are often confused with cell-wall deficient forms of other bacteria. After 1958 Gerlach worked at the Bavarian clinic of Josef Issels, who had pioneered the "whole-body" therapy program that included diet, exercise, vaccines, and other nonspecific interventions developed to bolster the immune system. While Gerlach was visiting Livingston and colleagues, they obtained samples of *Agrobacterium tumefaciens* from the Bronx Botanical Gardens. They then used the cultured bacteria to perform experiments with mice (Livingston 1984: 82).

In their travels to Europe, Livingston and colleagues had the opportunity to meet with Emmy Klieneberger-Nobel, who worked on the L-forms of bacteria at the Lister Institute, and with Ernest Brieger, who worked on filterable forms of the tubercle bacillus at Cambridge University. They also stopped to meet with Wilhelm von Brehmer and to learn about his dark-field techniques, that is, microscopes in which light is not directly transmitted through the specimen into the objective lens. However, reports of Livingston's work appeared in the American press, and a spokesperson for the New York Academy of Medicine discounted their claims. When Livingston and colleagues returned, they were in the middle of controversy (1984: 86–87).

In 1951 the Presbyterian Hospital of Newark and Memorial Sloan-Kettering Cancer Center in New York had each been awarded $750,000 in a grant from the Black-Stevenson Cancer Foundation. Livingston and colleagues had been expecting to begin work under their new grant when they returned from Europe. However, the story was to turn out differently:

> As Mr. Hardin, one of the directors of the Black grant, lay dying of cancer in the Memorial Center, he had been prevailed upon to sign a codicil to the

> bequest stating that we at the Presbyterian Hospital could not expend our share of the grant without the permission of Dr. Rhoads's Memorial Center. As it turned out, the only acquisitions that Dr. Rhoads would grant us were a new wing to be added to the hospital and the installation of a *high-voltage cobalt machine.* The sisters Black were betrayed as were Dr. Alexander-Jackson and myself, who had labored so long and diligently to establish a top-flight research laboratory devoted to the *biological* approach to the treatment of cancer, and *not* to radiation. It was our work that brought the $750,000 gift to Presbyterian in the first place, yet it was the machine this gift purchased that destroyed all that we had accomplished.
>
> At the time of the announcement of the Black grant, we were elated. We could foresee establishing preventive clinics across the nation that would screen patients and immunize them when they were bacteriologically positive, clinics that would promote better life habits, better nutrition, safer and cleaner surroundings, industrial and environmental control of carcinogens, earlier detection of precancerous lesions, and genetic counseling.
>
> It was a great dream while it lasted. (Livingston 1984: 88)

The situation continued to crumble. Their sponsor at Rutgers wanted to close down the laboratory. He offered to keep Livingston as an associate professor with a salary, but there was no room for Alexander-Jackson. In addition, Livingston reports that the Internal Revenue Service began to investigate her husband on the source of funds used for the European trip. She claims that she was told confidentially that "someone high up in New York in cancer" had spurred the investigation (Livingston 1972: 69). Although the claim may sound paranoid, other alternative-medical researchers have experienced similar suppression (Carter 1993).

Livingston and her husband moved to Southern California, him to be closer to business opportunities in Mexico and her to be closer to her family, which had moved to Los Angeles. A few years later her husband returned to New York, and shortly thereafter he died. She continued to work in a California clinic under very stressful conditions, and in 1957 she married a doctor there named A. M. Livingston. In 1958 she, her husband, the Dillers, and Alexander-Jackson attended the First International Congress for the Microbiology of Cancer and Leukemia in Antwerp. There they were able to talk with Gerlach and von Brehmer as well as other leading European researchers in the field. These included Ernest Villequez, director of the Central Blood Bank of France and Professor of Experimental Medicine at the University of Dijon; Nello Mori, director of the Instituto Microbiological Bella Vista in Naples; and Clara Fonti, president of the

Centro Internazionale Oncological di Viggio in Milan. They found many similarities of ideas, and in some cases some surprises. For example, Fonti had inoculated the skin on her chest between her breasts with bacteria (not cancer cells) cultured from human cancers, and this had produced a growth of basal cell epithelioma (Fonti 1958; Livingston 1972: 90).

Livingston was slowed by a heart attack in 1962, and Alexander-Jackson subsequently lost her job at Columbia, perhaps due to "high-level pressure" (Livingston 1972: 103). In 1968 the Livingston "vaccine" appeared on the ACS list of "unproven methods" (American Cancer Society 1968). There were, however, changes of fortune in the other direction as well. In the 1960s the work of British microbiologist Kenneth Bisset was becoming recognized in major journals.[34] He claimed that mycoplasma could break down into viruslike particles and that they and bacterial L-forms might play a role in malignant diseases. Thus, he helped pave the way for the increasing recognition of cell-wall deficient bacteria as "stealth pathogens," to use the phrase of microbiologist Lida Mattman (1993). At about this time the Livingston group also won a victory when, nearly twenty years after Diller had first attempted to organize the symposium at the New York Academy of Sciences, they finally gained approval to present their results before the academy. Livingston noted that she had over five-hundred requests for reprints based on the conference. Nevertheless, discussant Phyllis Pease, who worked in Bisset's laboratory at the University of Birmingham, commented that they faced an uphill battle: "As long as it is necessary to warn research workers that they may not mention endospores in their cultures or suggest that an apparent mycoplasma can be the L-form of a bacterium, for fear of ridicule or rejection, we are not likely to advance further in [a] line of research where these phenomena must be looked in the face and proved true or false on the evidence" (Pease 1970: 783–84).

An important addition at the 1969 meeting of the Academy was Florence Seibert, a senior microbiologist and biochemist who was best known for having developed the PPD (purified protein derivative) skin test for tuberculosis. She had studied the biochemical composition of extracts from rat sarcoma tumors, and she became interested in the microbiology of cancer in the 1950s when her friend and colleague Irene Diller showed her slides of acid-fast bacteria from her own tumor isolates. After a hiatus of several years following her retirement in Florida in 1958, Seibert opened a laboratory and became actively involved in research on bacteria and cancer. Although she managed to produce a series of articles on the topic, she did so in the face of funding problems. As she recounts,

> I soon found, like most of the other workers on this subject had already found, that the coffers of the wealthy organizations supported by funds given from the hearts of the crying victims for help, or by the funds supplied from our taxes, were empty when this phase of the work was requested. It was hard to understand, in view of the previous help I had had when I was working on the same subject but in the popular stream of thought. The challenge, however, became greater with every rebuff, as I remembered some of the history of science. Finally support did come from local bleeding hearts. (1968: 136)

She eventually did find funding, but all from local organizations and individuals.

The work became slightly less controversial, and received more recognition from mainstream researchers, when in 1972 Livingston found that her cancer microbe produced proteins similar in structure to human choriogonadotropin (hCG). This growth hormone is implicated in some cancers, and it may have a general importance as a tumor marker (Acevedo, Tong, and Hartsock 1995). Several other laboratories subsequently confirmed the finding, and their work in the late 1970s and into the 1980s brought the research into the era of molecular biology and contemporary immunology.

The other major development during the 1970s was the growth of the Livingston Foundation Medical Center, a clinic which Livingston and her husband opened in San Diego in 1969. In addition to using vaccines, they emphasized a diet high in abscisic acid that shared some key features with the Gerson diet. For the next twenty years Livingston continued her research and treatment of cancer patients, and she battled to protect the use of her vaccine. Ironically, as she pointed out in a talk given in 1989, she received official approval for use of the vaccine on farm animals for Marek's disease but not for use on humans. She also fought to get Medicare insurance coverage for her treatments, and, although she won the battle technically, the legal costs drained her resources (Livingston 1989).

In February, 1990, the California Department of Health Services issued a cease-and-desist order for her clinic to stop administering and prescribing the autogenous vaccine. Coincidentally, in the March/April issue of the American Cancer Society's journal *CA—A Journal for Clinicians,* the Livingston therapy was described under the heading of "Unproven Methods of Cancer Treatments." The notice stated that "Livingston has apparently mistaken several different types of bacteria, both rare and common, for a unique microbe" (American Cancer Society 1990). Although this claim was probably true as far as it went, it ignored a much more complicated research picture as well as the apparent successes with animal and human vaccines. The report also suggested that they had contacted a spokesperson for the state health department, who "agre[ed] that the clinic 'is probably in violation' of the 1959 California Cancer Act" (American Cancer Society 1990: 107). According to the current director of the clinic, with whom I spoke in 1995, there were apparently no complaints from patients that led to the state health department's decision, and the order seemed very odd to members of the clinic given the fact that autogenous vaccines are standard practice among allergists. Virginia Livingston died of heart failure a few months later, in June, 1990.

Enderlein, Von Brehmer, Issels, and the Nazis

The analysis so far has focused on North America, which has been where the bulk of the research advances were made during the decades after World War II. However, in order not to make overly hasty generalizations from the North American experience, it is helpful to understand the history of research on bacteria as etiological agents of cancer that took place in other countries. During this time period, the other world center of biomedical research was Europe. Until the Nazis came to power, Germany had the most advanced scientific research community in the world, and France also had a long-standing microbiological research tradition. In Britain some researchers advocated the bacterial etiology theory—such as William Russell, James Young, and William Crofton—and as in the United States they were attacked or ignored. The discussion here will be limited to Germany and France, which provide a stronger comparative perspective than the British case both in terms of institutional patterns and in terms of the types of bacterial theory.[35]

Germany provided the model for scientific medicine in the United States, including educational experiments such as those at the Johns Hopkins Medical School, and it also provided some of the most advanced research on pleomorphic microbiology. Probably the most influential German researcher on pleomorphism was Günther Enderlein, who began as a student of the botanist Wilhelm Pfeffer and started work at the Agricultural University in Berlin. Officially a zoologist, during World War I he worked as a serologist and doctor in the army due to the shortage of physicians. His theories of a pleomorphic bacterium or fungus emerged from this experience, and in 1916 he began submitting reports based on his studies of typhus. That research became the basis for his 1925 book *Bakterien Cyclogenie.* Enderlein was therefore endorsing a cyclogenic theory of bacterial pleomorphism at about the same time as Glover was. What distinguished Enderlein's work was his focus on blood-borne microorganisms. In general the continental researchers tended to emphasize an endogenous blood-borne microbe rather than a latent intracellular tissue pathogen.

As early as 1901 the German researcher Otto Schmidt had found microbes in the blood of cancer patients (Enby 1990). Blood-borne pathogens are recognized causes of diseases in microbiology today. For example, Lida Mattman writes, "Diseased lymph glands, especially if in the abdominal cavity, feed diptheroids and other pleomorphic organisms into the bloodstream. Blood cultures may give a diagnostic clue in those conditions" (1993: 312, citing Fleisher 1952). Thus, what makes Enderlein's theory controversial is not the idea that bacteria, viruses, or fungi can appear in the blood of sick people, but that they are part of a microbial cycle even in healthy people.

Enderlein's basic argument was that the blood, like the digestive tract, has a native flora, and the blood-based microorganisms play an essential role in the life process. However, when the body's internal environment changes and the blood pH changes—due to poor nutrition or exposure to environmental stressors—the blood-based microbes transform into higher stages (up to the level of fungi) and cause disease. Enderlein called his highly pleomorphic microorganism *Mucor racemosus fresen,* and he believed it has a life cycle of fourteen phases. At its lowest stage, the microorganism exists as a "protit," a living, nonmoving, apathogenic, protein particle about the size of .01 micron and similar to what today would be called prions (Enby 1990: 28). The second stage takes the form of tiny protein balls composed of a number of protits. At the tenth stage the organism reaches a "multinuclear" form that breaks off from the outer membrane of red blood cells to become independent "bacteria." By the thirteenth phase the organism has reached a fungal form, and in the fourteenth phase it produces spores. If this microorganism exists, it would be quite different from ordinary fungi or even the mycobacterium, which constitute a bridge group between the bacteria and fungi, because known fungi and bacteria do not show such a range of pleomorphism.

Enderlein did not work alone or in secret; rather, he found significant support for his ideas. During the 1930s he became director of production control of the pharmaceutical company Sanum. Their first product was Utilin, an antituberculosis drug developed by the doctor Friedrich Friedmann and derived from microorganisms found in the blood of a sea turtle. (It is now used for AIDS patients.) Friedmann's treatments had already received international recognition, and prior to World War I he had lectured before the U.S. Senate and published a document with the U.S. Government Printing Office (Friedmann 1913). During the 1930s, another German doctor, Wilhelm von Brehmer, extended Enderlein's work by developing a related theory of cancer based on his observations of the fungus *Mucor racemosus* in the blood.

The claims of Enderlein and other Germans who studied pleomorphic blood-borne microorganisms were heterodox enough, but according to the Swedish doctor and Enderlein biographer Erik Enby, when the Nazis came to power in 1933 Enderlein's ideas came into conflict with the Nazi doctrine of purity of blood. Enderlein suffered severe criticism from the Nazi doctors, and his fellow researchers Friedmann (who was Jewish) and von Brehmer were openly persecuted. Prior to the Nazi rise to power, von Brehmer had sued Hitler for defamation of character, and he was an outspoken critic of the leader. When later pressed to join the Nazi party, he resigned his position. Enby writes that von Brehmer only survived the Nazi years due to his reputation as a scientist and his connections with people high up in the power structure. Regarding Friedmann, Enby writes:

> [T]he biological treatment was viewed by Nazi doctors as direct competition for the chemotherapy industry. Directors of numerous German tuberculosis sanatoriums also harshly criticized the Utilin vaccine, which posed an economic threat to their clinics. . . . Under such severe pressure, Dr. Friedmann fled to Monaco in 1937. (Enby 1990: 8)

After the war, the endogenous blood microbe theory remained unpopular, particularly among the former Nazi doctors who continued to practice and influence German medicine. However, Enderlein and colleagues continued to do their research, and versions of their biological products continued to be manufactured.

Former Nazi doctors were also partly responsible for the suppression that the German doctor Josef Issels suffered during the decades after World War II. In 1951 Issels founded the first hospital in Germany that pioneered a multimethod, nontoxic treatment of cancer. He advocated a multicausal theory of the etiology of cancer that included heredity, lifestyle, exposure to carcinogens, and microorganisms (Issels 1975). He also employed Franz Gerlach, the Austrian scientist who maintained ties with Livingston and the network of North American researchers. Gerlach (1948) researched and published on the microbial theory of cancer, and he helped produce vaccines that Issels used as part of his multimethod therapy. Although Gerlach's research and vaccines were probably not the main cause of the legal problems that Issels encountered, they contributed to his controversial status in the German medical community.

In the 1960s Issels fought and won a long legal battle against false charges of manslaughter and fraud that his biographer Gordon Thomas (1975) argues were brought about by the German Cancer Society and the Bavarian Medical Association. Issels's license was never revoked, and after the hospital closed in 1973 he continued to treat patients on an outpatient basis until his retirement in 1987. From 1981 to 1987 he served as an expert member of the German federal government commission in the fight against cancer. He retired to Florida, where there is a clinic that uses some Issels-based methods, but within the confines of American law, to treat patients of chronic diseases. In 1996 he accepted an affiliation with the Gerson Research Organization, and this new relationship promises to bring his insights to the clinical practice of the affiliated Centro Hospitalario Internacional Pacifico, SA, in Tijuana. I spoke with Issels briefly at the meeting of the Cancer Control Society in 1996, and he reiterated his belief in the importance of mycoplasma as contributing agents in cancer pathology.

French Mandarins

In France research on the possible etiological role of bacteria in cancer dated back to the late nineteenth century. Probably the most well-known early work is that of Eugène Doyen, a prominent French surgeon at the Hospital of Paris. At the turn of the century Doyen defended his research on the microbial theory and the efficacy of a serum before a committee drawn from the Pasteur Institute and the Société de Chirurgie. Although the committee turned in a positive report on his work on the organism that he named *Micrococcus neoformans,* researchers in Britain criticized Doyen's work severely and ultimately it was rejected (Boesch 1960). Several other French researchers supported the bacterial theory, including Gustave Rappin (1939), the director of the Pasteur Institute of Nantes, who for more than a half century studied "the presence of granular forms of a microbial nature in cancerous cells" (Villequez 1969: 26).

This section will focus on more recent examples of French research guided by the general microbial theory of cancer: the work of Ernest

Villequez and Gaston Naessens. The Dijon-based professor and medical researcher Ernest Villequez has already been mentioned as integrated into the post–World War II international research network. His work is particularly interesting from a social-science perspective because he left behind a record of the intellectual suppression that he faced. In his book *Human Cancer: The Forbidden Study,* published in the wake of the events of 1968, Villequez not only defended his research but also analyzed and condemned the cancer research community in his country. Unlike similar critiques of cancer research in the United States, which generally have focused on corruption by economic interests, Villequez focused his critique on the centralized nature of the French research system and the stifling conservativism of the "mandarins."[36]

According to Villequez, the end of the nineteenth century was a "period of great hopes placed on the possible proof of a microbial origin of cancer" (1969: 25). He argues that medical consensus shifted away from the bacterial etiology because the numerous late-nineteenth-century studies had found diverse microorganisms (bacilli, cocci, or microscopic fungi). As he comments, "These disparate and contradictory observations suggested that the results were the product of accidental contaminations" (35). The renewed acceptance of bacterial pleomorphism after the turn of the century made it easier to reconcile these diverse observations with a theory of microbial etiology, but by then many cancer researchers had closed the door to the theory. Furthermore, after Japanese scientists demonstrated the role of chemicals as cancer-causing agents in 1916, opinion shifted even further away from microbial research, with the exception of the occasional studies on viruses (26). Another factor that Villequez argues "contributed to the systematic demise of the infectious theory" was the creation of the research institutes at the beginning of the century (35). In the institutes, "The training of the leaders of oncology did not involve laboratory studies in infectious pathology, which meant that without a doubt they were deflected from entering into a strange terrain that they found difficult to exploit" (35).

The resulting climate of opinion in France was, according to Villequez, the "dogma" of the bacterial sterility of the blood and of tumors among French cancer researchers. "This perfect academic and professional ignorance leads to a prohibition (to avoid losing time) of studies and publications contrary to the orthodox teaching" (1969: 39). Villequez adds that on repeated occasions hematologists gave his presentations a stony reception. "My presentations were merely morphological; in other words they were limited simply to visible things that hadn't been seen before, but it appeared so fantastic to them that I might as well have been describing the details of the flora of another planet" (40). Editors of journals also refused to publish his work over a period of three decades, and conferences that he and colleagues held on the microbiology of cancer in 1958, 1959, and 1960 in Belgium and France resulted in a situation in which "none of the invited medical figures attended the assemblies" (15). As a result, their efforts were abandoned (123).

French medical opinion also hardened as a result of two very visible events: the Lorenz and Naessens trials. F. W. Lorenz was the secretary of von Brehmer who became his student. When the town where they worked, Badkreuznach, fell under French occupation at the end of the war, Lorenz moved to Paris, took with him some bacterial samples from von Brehmer's laboratory, and used those samples to make a cancer vaccine. As Villequez writes, "Von Brehmer prosecuted Lorenz through the court of the Seine not only out of anger . . . but also because he hoped that the trial would involve French scientific interest and therefore awaken interest in his work" (1969: 46). In August, 1948, a committee of experts was designated, and in 1951 they made public their negative conclusions: 1) that *Siphonospora polymorpha* (von Brehmer's name for the cancer-causing microorganism) was produced by a "method of coloration that is personal to the research and that produces the appearance of little motile particles"; 2) that "if *Siphonosopora polymorpha* existed, it would have been huge in the electron microscope that was used"; 3) "the cultures of *Siphonospora polymorpha* are merely common air-borne germs" (47). A press release from the Ministry of Health informed the public that the affair was closed (48). M.-E. Richier-Chevrel (1951), a medical student in Paris, defended a thesis on Lorenz and on the need for tougher laws in France to combat charlatanism (49). Because Lorenz was neither a medical doctor nor a scientist, but he claimed the title of medical doctor when he came to Paris, he was clearly engaged in fraudulent activity (50). However, von Brehmer also became associated with the charlatanism, particularly in press accounts of the "phantom bacteria" (50). In reply to criticisms of his work, von Brehmer wrote a letter to the editor of *Ce Matin* in 1952, but the letter went unpublished and his work rapidly passed into obscurity after his death in 1958 (45, 50).

Villequez contends that there were technical errors in the work of both the expert committee and the thesis. In December, 1951, he met with one of the experts and explained the techniques used to culture pleomorphic bacteria, and he concluded that the studies were conducted in a manner that was neither rigorous enough to avoid contamination nor methodologically appropriate for the study of pleomorphic bacteria (1969: 48). Regarding the thesis he pointed out some technical errors that the student made. These included the failure to maintain rigorously sterile procedures when drawing blood and also the failure to maintain rigorously anaerobic conditions in the bacterial cultures (52). Furthermore, he noted that even if the germs were not due to contamination, the fact that they were common germs did not mean that they had no role in the cancer process. The next question he asked would turn out to be central for the survival of the bacterial etiology theory. "Why," he asked, "can't a common germ, in certain species of its group and under certain conditions, become an agent of carcinogenesis? Cancer research suggests that one accepts the postulate of the intervention of diffuse parasites that exist in the organism in a latent state" (52). Although this argument may seem like an auxiliary hypothesis maneuver, it will prove central in the subsequent molecular biology studies of bacteria from cancer samples.

In the generation subsequent to Villequez, Gaston Naessens is probably the best-known French researcher to support the theory that a pleomorphic microorganism plays a role in the etiology of cancer. Naessens grew up near Lille, a northern city that is known in the history of microbiology as the place where Antoine Béchamp taught. Béchamp was an advocate of an extreme form of bacterial pleomorphism, and although Naessens did not become aware of Béchamp's work until later in life, he probably was exposed to that line of thinking in his college studies in science at the University of Lille. During the Nazi occupation, Naessens continued his college education in southern France among displaced professors from Lille. Born into a family of bankers, he had the luxury to pursue his inventions. Having worked in a laboratory for blood analysis, he became interested in blood parasitology, and he invented a darkfield microscope that used prisms and laser-beams to obtain high levels of magnification. Although the technology is different from that of the Rife microscope, Naessens also claimed to achieve high levels of magnification and resolution. The higher levels of magnification and the darkfield technology allowed him to observe the smaller, viruslike phase of what he claims is a sixteen-stage, pleomorphic microbial blood cycle. Like Villequez and the Germans, he saw the microbe as part of the natural flora of the blood that probably play an essential role in growth and reproduction, perhaps by producing necessary growth hormones. Naessens believes that when the immune system is compromised and "blood inhibitors" are not present, the microbe can go into extended, funguslike phases. When I interviewed him at his home in Quebec, he explained that he thinks of the "somatid" cycle as primarily diagnostic of chronic or advanced diseases, but he also believes that in the advanced stages of the somatid cycle excess growth hormones are released that may contribute to cancer.

Although Naessens today is aware of the entire research tradition outlined in this chapter, according to Villequez he was a novice when in 1960 he attended the International Congress of Microbiologists and Clinicians of Cancer and Leukemia, for which Villequez was the keynote speaker (Villequez 1969: 53, 104). He apparently disagreed with Naessens on some aspects of his research; for example, Villequez argued that the microphotographs made by Naessens involved misinterpretations of fungal filaments (58). Nevertheless, Villequez testified in favor of the bacterial theory of cancer at Naessens's trial in the early 1960s (58).

Naessens was tried for his use of Anablast, an anticancer serum that, like the Glover and Doyen sera, was developed from a horse injected with bacteria cultured from the blood of cancer patients (Villequez 1969: 53).[37] Years later patients who attributed their cure to Anablast came to his defense. However, Naessens was not a doctor, nor did he report his results in a medical journal. He therefore occupied a position as an outsider to the medical establishment and was particularly vulnerable to suppression. The result of the trial for illegal exercise of medicine and pharmacy was that Naessens was fined, his laboratory was closed, and much of his equipment (but not his microscope) was confiscated.

To pursue his work without interference, Naessens moved to Corsica. However, word spread of his whereabouts and soon he was invaded by hundreds of patients. The French medical authorities began an investigation that led to another trial in 1965. Meanwhile, he had relocated to Quebec, where he developed another anticancer drug, 714-X, the one that is now best known in the alternative cancer therapies movement today and to some extent in the AIDS movement. The drug affixes nitrogen atoms to camphor molecules. Naessens reasons that because cancer cells are nitrogen traps and because camphor is attracted to cancer cells, the drug provides the cancer cells with the necessary nitrogen that would be robbed from healthy cells. In the process, the immune system is able to recover to the point where it can then destroy the cancer cells. The toxicity of camphor may also be a contributing factor.

In 1985, the Quebec Medical Corporation prodded the government into indicting Naessens, and he went to trial in 1989. The charges included murder, and the potential sentence was life imprisonment. He had treated a terminally ill patient who had asked him to treat her, but according to Naessens her cancer was too advanced and there was not enough time for his treatment to work. (The trial is described in detail in *The Persecution and Trial of Gaston Naessens,* by journalist/writer Christopher Bird.) In a surprise event for alternative cancer therapies, Naessens won the trial. The prosecution brought in male, scientific experts who questioned Naessens's theory of a pleomorphic microbial cycle in the human blood. The defense focused on the therapy rather than the theory, and it brought as witnesses men and women from all walks of life, including some leaders of Quebec society. Many of the witnesses for the defense stated that they owed their life to Naessens, and the jury swung in their favor. Although Naessens is now very careful not to prescribe or administer 714-X, the Canadian government allows the drug to be administered under special permission. Requests for the drug have been high during the 1990s, and Naessens's laboratory is apparently busy with its ongoing work of blood analysis.

Concluding Comments

By examining the most well-known cases in several countries, it is now possible to draw some conclusions about the history of suppression of bacterial theories and therapies for cancer. First, the details of the cases suggest that the experience of the researchers and clinicians was by no means uniform. In Brian Martin's terms, there is a sliding scale of events between intellectual suppression and repression, that is, between marginalizing techniques and the use of extralegal violence (Martin et al. 1986: 2–3). In general, researchers tended to experience milder problems of suppression such as overcoming gatekeeping hurdles or loss of academic positions. In contrast, those who went on to produce and use clinical products faced legal challenges and, in some cases, extralegal repression. In Nazi Germany, where the legal system became subverted to a dictatorship, the attacks on the unorthodox clinicians did not need to remain within the legal system, but in the United States there were also instances of apparently extralegal repression, such as in the Rife case. In other words, there was a range of responses to the bacterial etiology theory in its diverse variants: the initial enthusiasm that greeted the Glover serum, the gatekeeping and cognitive cronyism that was Villequez's major complaint, the bureaucratic roadblocks that Glover experienced, the diversion of funding that Livingston experienced, legal proceedings that in some cases represented a legitimate state interest in preventing the illegal practice of medicine (as in the Lorenz case), more obviously rigged legal prosecution in which interested segments of the society subverted the principle of equal justice (as in the Rife case), and cases of extralegal repression, such as the claims of violent repression in the Rife case.

For those familiar with the history of alternative medicine, the range of suppression and repression that I am describing will come as no surprise. The contemporary conflicts have their parallels in the struggles between the allopaths and empiricks in the nineteenth century (Coulter 1987). In the twentieth century, the conflicts continued in the form of professionalized, state-sanctioned medical societies versus a wide variety of alternative medical practitioners. In some cases, such as osteopathy, the resolution of the conflict has been through incorporation, whereas in others it has been through legal recognition of a parallel practice, as in chiropractic (Baer 1987, 1989). The problem that now emerges is to develop a sophisticated explanation of why the nonviral microbial theories and therapies of cancer met with such hostility, rather than being simply ignored, enthusiastically accepted, or cut down to size and incorporated as occurred with viral oncology.

3

Culture and Power in Cancer Research

Reconstructing Interest Theory

What is a good explanation of the pattern of suppression that occurred for those scientists and clinicians who advocated the microbial approach to cancer? Within alternative medical circles today the standard explanation is that the emerging cancer establishment was corrupted by financial interests. This approach is applied generally to explain a wide range of suppression of alternative cancer therapies, and it has a strong rhetorical appeal. By pointing to the financial interests that shape the cancer establishment, critics of the medical-pharmaceutical complex paint alternative medicine advocates as the rational, heroic underdogs in the struggle for truth, justice, and the very American value of medical freedom. In contrast, members of the cancer establishment become at best cogs in a machinery they do not even understand and at worst part of a more-or-less conscious conspiracy of powerful, material interests. As I have shown in previous research, pinning material interest motives on one's opponents can be a source of great rhetorical power in public debates (Hess 1993). The power is so great that both sides tend to try this strategy. Thus, rhetorics of money-grubbing doctors get paired off against those of money-grubbing quacks, and explanations of why the suppression occurred become mere caricatures of good social science.

Where does one look for an alternative explanatory framework that does not jettison the basic insight that material interests were involved? In the interdisciplinary science and technology studies (STS) literature, interest theory was a popular approach for the social studies of scientific controversies during the late 1970s. However, the theory fell out of favor in the early 1980s after a series of very sharp criticisms.[1] The intellectual justification for the abandonment of interest theory is that it tends to flatten scientists into unidimensional beings; in other words, scientists are portrayed as "interest dopes." Although the criticisms of interest theory pointed to some serious problems that needed to be addressed, the proposed alternative viewed interests as something that scientists produced by making others "interested" in their work (e.g., Callon and Law 1982; Latour 1988). Unfortunately, this alternative tends to invert the picture by turning interested social groups into scientist dopes. The alternative therefore loses the original insight that powerful professional and financial groups can play an active role in restricting which knowledges make it onto the playing fields of consensus science. The history of alternative cancer theories and therapies represents a good reason for revisiting interest theory, because it is difficult to write that history without referring to the active role that interests play in the suppression of some researchers and therapies. To my mind, the solution to the problem with interest theory is not to dismiss economic and professional interests as shaping variables in the explanation of scientific controversies, but to find a better way of thinking about interests.

Thus, I advocate neither the widespread and almost unquestioned use of interests explanations that appears sometimes in the alternative cancer literature nor the opposite extreme of much of science-and-technology studies, which after the early 1980s tended to drop interest theory. This chapter therefore has a double import: expanding the sociocultural understanding of the politics of cancer research, and showing how the culture concept can make interests explanations more sophisticated. My goal is to explain the suppression of the microbial theories, but not by reducing the medical profession to a sociological puppet. Rather, I will show how scientists, the medical profession, industrial groups, and so on define their interests in diverse intellectual, professional, gender, national, and scientific cultures.

The Culture of Orthodox Cancer Research

A sophisticated use of interest theory would be directed toward explaining not individual historical cases of intellectual suppression, but instead the fact that an entire research tradition became so heterodox that its advocates faced severe suppression. In other words, the larger problem is to explain the historical emergence of the general contours of a cancer research culture. In the United States, this culture congealed around key people and institutions, such as James Ewing and Cornelius Rhoads of what is today called the Memorial Sloan-Kettering Cancer Center, Francis Wood of

Columbia, Morris Fishbein of the AMA, and Howard Kelly of Johns Hopkins. By the 1930s and 1940s the consensus became institutionalized through organizations such as the National Cancer Institute and the American Cancer Society. Ralph Moss (1989), who develops a sophisticated use of interest theory, describes this confluence of people and institutions as the "the cancer industry": the American Cancer Society, the National Cancer Institute, the major cancer hospitals, pharmaceutical companies, and the medical profession.

At its inception in the 1930s and 1940s, the National Cancer Institute began directing research away from the infectious theory in general and bacterial research in particular. According to STS analysts Kenneth Studer and Daryl Chubin, when the NCI was formed the surgeon general appointed a committee to outline the type of research that the new institute would be doing. The committee resolved that mammalian cancer was not infectious and that viruses were among the etiological agents that could be disregarded. As Studer and Chubin comment, the NCI founders "knew what they did not like—viruses—and this no doubt influenced, through funding of research at major universities, the energy expended on the discovery of new oncogenic viruses" (1980: 25).

Studer and Chubin suggest that the founding of the NCI was an important factor in the "great slowdown" of virus research that occurred during this period (1980: 21). This view is echoed in a comment from Ludwik Gross, a leading tumor virologist: "In the 1930s and 1940s a young scientist applying for a grant to carry out a research project in the field of tumor-inducing viruses was faced with considerable difficulties in obtaining a place to work or sufficient funds to carry out his studies" (1983: 1086). Likewise, medical historian James Patterson adds that the six-member National Advisory Cancer Council of the NCI—the group that gave Glover so much trouble with his publication plans—included major cancer insiders such as James Ewing, who pushed research toward major centers, including Memorial Hospital (1987: 131–35). Patterson notes that tumor virologist Peyton Rous complained about the in-group that dominated the NCI, and, "observing the interest in the NCI of drug companies and of manufacturers of X-ray machines, he noted that 'the guns of the powerful laboratories run by the electric and engineering companies at once began to say, "Boom! Boom!' "" (1987: 130). The council also included Wood of Columbia, who worked to discredit Glover and wrote of bacterial theories that they had "always been the darling of crackpots and dabblers" (98).

Another organization that became an integral part of the cancer establishment was the American Cancer Society (1966, 1990). Although Livingston originally received some support from the organization, her therapy ended up on its list of unconventional cancer therapies, as did many other alternative therapies. The ACS is merely a private organization and it has no official mandate from the U.S. government or the American people, but still its positions have had a great impact on the federal government. As Moss (1989) documented, therapies that end up on its list of unconventional therapies usually cannot get approval from the Food and Drug Administration and instead suffer FDA suppression. Furthermore, the research priorities of the ACS have an impact on the much larger funding flows from the NCI, so much so that the ACS has sometimes been called the "tail that wags the dog." Economists James Bennett and Thomas DiLorenzo (1994) demonstrated that the ACS and other large health charities tend to fund within the networks associated with their boards; in the case of the ACS the board includes many members from the major cancer research organizations. The in-group nature of the funding process and peer review for cancer research makes it difficult for outsiders to break into the system.

To summarize, by the middle decades of the twentieth century in the United States a cancer research community emerged within the larger arena of the medical profession and biomedical research science. This cancer research community congealed under the leadership of a fairly close-knit "cancer establishment" that by the 1950s and 1960s included the major research hospitals, the NCI, and the ACS, with support from the FDA. The establishment cancer community maintained that cancer was not an infectious disease, at least for the vast majority of human cancers. Rather, cancer arose from a variety of other factors, of which the first to be documented were hereditary predisposition and chemical carcinogens.

The consensus of the cancer research community in the United States took two forms: a consensus on etiology and a consensus on therapy. At the 1926 conference in Lake Mohonk, New York—the one that Coley attended and described—the major causes of cancer being considered were local trauma, heredity, and chronic irritation. In other words, by the 1920s the infectious theory had already been marginalized. By the 1980s and 1990s heredity continued to occupy an important place in the consensus knowledge of cancer etiology, but environmental factors such as chronic irritation and local trauma had given way to diet, smoking, and sexual/reproductive behavior as major environmental and lifestyle causes of cancer.

Although there was a lack of support for viral oncology during the formative decades of the NCI, viral oncologists slowly amassed a large number of studies that linked some animal cancers to viruses. In the 1960s, after the development of the polio vaccines, viral research returned as a priority in the NCI research agenda (Wade 1971). However, by this point viruses were viewed as possible etiological agents only for a few types of human cancer. Furthermore, the NCI's Virus Cancer Program became controversial by the early 1970s, in part due to its exclusive club quality (Rettig 1977; Culliton 1974). As STS analyst Joan Fujimura (1995) argues, the proto-oncogene theory provided a justification for expenditures on the viral research program, because viral oncogenes not associated with human cancers were shown to have homologs in human proto-oncogenes. The proto-oncogene theory therefore moved viral oncology into the center of consensus cancer research, albeit under the encompassing wing of a general molecular-biology research program.

On the therapeutic side, a similar consensus emerged in the cancer research community that developed from the early twentieth century emphasis on surgery with adjuvant radiotherapy. As occurred with theories of the cause of cancer, by the 1980s and 1990s the list of accepted therapies among mainstream cancer researchers in the United States was considerably more diverse. Chemotherapy and hormonal therapy had come to occupy an important part of mainstream cancer therapies, and more recently hyperthermia and specific immunotherapies achieved some acceptance. Even some nonspecific immunotherapies that used bacteria, such as Coley's toxins or the BCG tuberculosis vaccine, had achieved a modicum of recognition among immunotherapists and some limited testing. However, therapies that were explicitly based on the bacterial-etiology theory, such as Glover's serum and Livingston's vaccines, were not even within the horizon of consideration, and many other alternatives—including several nutritional and metabolic therapies with some experimental and clinical data to support them—continued to be suppressed.

This brief sketch is sufficient to make the point that mainstream cancer research is not a monolith. Its theories and therapies changed substantially over time amid numerous and intense internal controversies. For example, for a long time smoking was not considered an important cause of cancer, and today a similar debate has erupted over epidemiological studies that suggest that chemical carcinogens and pollution have been greatly underestimated as risk factors. Yet beyond the internal controversies there is an unquestioned set of assumptions—the doxa or, in the terms of philosopher Imre Lakatos (1978), the hard core of the research program—that includes the theory that human cancer is, for the most part, not an infectious disease.

Stated positively, cancer research today has increasingly come under the encompassing wing of the molecular-biology research program. From this perspective, cancer is the outcome of a mutagenic process based on multiple hits to the DNA, a process that allows research programs on heredity, viruses, and chemical carcinogens to coexist peacefully amid research on other promoting factors such as hormones. Within this set of assumptions, viruses are seen as relatively unimportant for the major human cancers, and bacteria are seen only as opportunistic infections that play no role in tumor genesis or promotion. Together, these assumptions form the intellectual core of the modern culture of cancer research.

To describe mainstream cancer research as having a culture does not imply that the culture is an unscientific one that cannot be justified with sound arguments. As advocates of the infectious theory of cancer realized, they suffered from studies in the 1890s and into the first years of the twentieth century that unsuccessfully attempted to culture bacteria from malignant tissues (Glover and Engle 1938: 9; Villequez 1969). Furthermore, the studies that did culture bacteria or fungi did not isolate a specific, unique organism as in the case of other bacterial diseases, and the techniques required for successfully culturing the purported cancer microorganisms were not widely enough diffused to make replication easy. Therefore, the theory that bacterial cultures from cancer tissues were only contaminants was a reasonable interpretation of the inconsistencies of the data.

Furthermore, during the first decades of the twentieth century, research pointed to the noninfectious causes of cancer. For centuries the correlation of scrotal cancer and chimney sweeps had been known, and thus chemical carcinogenesis was by no means a new idea. A few years after the turn of the century the theory received new impetus when Tokyo University researchers Katsusaburo Yamagiwa and Koichi Ichikawa (1916) induced cancer in rabbits by applying coal tar. In the next decade Ernest Kennaway (1955) at the Royal Cancer Hospital in London produced the first chemical carcinogens and identified the first chemical carcinogen: 3,4-benzyrene. Another major development was X-ray and radium therapy, which came into widespread use during the first decades of the twentieth century and resulted in many cases of radiation-induced cancers among health-care professionals. By the 1920s reports were also showing that patients who received radiation treatment and workers who painted radiation dials on clocks and watches were suffering from cancers related to their exposure to X-rays and radium. Yet another development by the 1920s was the experimental research on cancer in mice, which provided convincing evidence for hereditary factors in the etiology of cancer. Thus, a reasonable scientist at that time could defend a position that cancer was not an infectious disease.[2]

Conversely, a reasonable scientist who defended infectious theories of cancer could also argue that irritants such as coal tar made it possible for latent viruses or other microorganisms to become active (Gye 1925a: 116). In other words, sound arguments could also be found to justify the theory that cancer may be an infectious disease. Roughly contemporary with the coal-tar studies were the first documented viral studies of cancer. In the first decade of the twentieth century scientists found a virus that induced a type of leukemia in chickens, and within the next three years this research was followed by the work of Peyton Rous and Japanese scientists, who demonstrated that viruses could cause sarcomas in fowls.[3] The early research on viruses focused on nonmammalian vertebrates, and this accounts for some of the lack of interest in viral research during the first decades of the twentieth century. However, the evidence continued to mount over the decades, and some tumor virologists continued to defend the theory that many human cancers have a viral origin and may be passed on by vertical transmission from parents to offspring (Gross 1983: 1003).

Although the viral studies were keeping the door open to infectious theories of cancer, there is also some evidence that the virologists were drawing boundaries of their own as their field was constituted. At the 1925 annual meeting of the British Medical Association, pathologist William Gye led a discussion of his research on cancer and filter-passing viruses that carefully ruled out the bacterial theory (Gye 1925a, 1925b). Although he noted that cocci were usually present, he referred to them as "contaminating" (1925b : 189). Regarding the bacterial theory, he wrote:

> Contaminations have been interpreted differently, however, by other authors; to some, especially in the field of cancer research, they represent an interesting phenomenon. The filterable virus is regarded as merely a phase in the life-history of the organism, the contaminant constituting a second phase. Taking this view and disregarding classical doctrine of the fixity of form of bacteria—or at least straining beyond justification the fact that variations in size and form occur—such authors are not deterred from lumping together as one organism a mixture of virus, cocci, and bacilli. The view seems to me to be fantastic. (1925b: 191)

William Crofton, a lecturer in pathology at University College, Dublin, criticized that interpretation and the empirical claim that the tumor virologists had been unable to culture bacteria from cancer samples. Convinced that research on the bacterial etiology theory was not getting a fair hearing, Crofton performed experiments that he believed replicated the work. However, he could not interest even the tumor virologists. As he wrote:

> But in spite of the complete satisfaction of Koch's postulates in these cases, this work has had no better result on the minds of such workers as Murray, Gye, or Lumsden than the original work of the Leytons. It is very unfortunate, and very much to be deprecated. Gye—I should have thought—would be interested, but he is obsessed with the virus theory of the workers for the Medical Research Council—a sacrosanct theory to question which appears to be considered blasphemy. (1936: 117)

This passage, together with correspondence from Gye to Coley that attempted to dissuade Coley from the general microbial theory, suggests that Gye and other tumor virologists were hostile to the bacterial theory. However, there may have been exceptions. For example, recall that Livingston noted that when she discussed her research with Rous, he was interested and, when they reported that they had grown the Rous agent in artificial media outside the living cell, the tumor virologist said "he did not think this was unlikely or impossible" (Livingston 1984: 79). Thus, there was room for a range of opinions among tumor virologists, and some may have been relatively open to bacterial and fungal research. Nevertheless, the virus and the bacteria researchers were not necessarily allies. As Crofton's comments on Gye suggest, the tumor virologists may have even seen the bacterial researchers as competitors who were discrediting their already controversial position.

Coley suggested a very different course for cancer research than the lines that it was taking in the 1920s and 1930s. Regarding the Lake Mohonk conference in 1926, he wrote, "It was curious to note that on the very page of one of the leading daily newspapers containing reports of the papers of Dr. Wood and Dr. Roussy in which they emphasized that, in their opinion, cancer cannot be due to a germ or parasite, there appeared an article entitled: 'Cancer Parasites Bred and Observed under Microscope' " (1926: 224). Coley then described the work of scientists in Berlin who reported before a German scientific congress that they had isolated a cancer microbe and had used it to produce cancer in animals. Regarding the research program that he saw being ruled out at the Lake Mohonk conference, Coley wrote, "Until it is settled beyond the shadow of doubt that cancer is not due to a microorganism, we believe that every effort should be made to stimulate to the utmost cancer research along these lines rather than to attempt to hinder or to discredit it" (1926: 224). He then warned, in passages that could be spoken today with little change, that there was dangerously slow progress in the control of cancer and an apparent increase in cancer rates.

Foreseeing epidemiological advances that were not achieved until decades later, Coley also called on the American Cancer Society (then the American Society for the Control of Cancer) to conduct comparative epidemiological studies, particularly of diet. He noted that South Asians had relatively low rates of cancer but that these rates increased when they adopted the habits and diet of the English, and he suggested that there might be some biochemical factor in the diet that "would be of great value in the prevention if not the cure of the disease" (225). In other papers published on the "problem of cancer control," Coley also defended the microbial theory and the need for more research on epidemiology and diet (Coley 1928, 1931).

Why did researchers and organizations not follow Coley's suggestions on possible dietary and infectious etiologies? Why did they not follow up on the many reports during the first decades of the twentieth century that suggested a viral or bacterial etiology, or both? To argue that the early cancer researchers could justify their choice of a noninfectious theory with sound arguments misses the point; they could have also justified a choice in favor of more research under a viral, bacterial, or general microbial theory as well as therapeutic studies of vaccines and diet. One might argue that the advocates of the infectious theories were poorly organized, bad at enrolling allies, and, to use everyday parlance, just plain lousy at marketing their research. They should have been able to interest the leaders of the medical profession and capture the emerging cancer industry. They were merely bad scientific entrepreneurs; if only they had studied Pasteur more carefully, they might have learned the valuable lessons of scientific entrepreneurship and won the case against colleagues such as Ewing. However, this argument leads to a blame-the-victim type of history that celebrates the superior skills of the winners. Can we not find a more sociological explanation of the failure of the infectious theories, particularly its bacterial variants?

Certainly, considerations of evidence—such as the difficulty of culturing consistently the same microbe from cancer samples, the undefined nature of viruses during the first decades of the twentieth century, the lack of general knowledge on bacterial variation, and the limitations of microbiological technology in an era prior to electron microscopy and molecular biology—provide some causal factors that shaped the failure of infectious theories. However, it is clear that for some serious researchers there was not enough evidence to abandon infectious theories of cancer at the time. Yet, opinion in the cancer research community solidified rapidly in the face of a great deal of interpretive flexibility in the evidence. Without ignoring the power of evidence to change minds in science (which is, of course, the attraction of science as a way of life), a complete account of the emergence of a consensus of cancer research against infectious theories needs to go beyond the issues of evidence to social factors that also shaped the direction of the emerging consensus of cancer research. Interest theory provides a starting point, but as we shall see it is only a starting point.

Interest Theory, Version I

The alternative medicine literature often suggests that the basic contours of orthodox cancer research were shaped by outside industrial interests, that is, interests exterior to the medical profession. For the North American case, and to some extent for Britain and other countries, the category "industrial interests" could be operationalized as the financial interests of the Rockefeller empire and affiliated empires such as that of Andrew Carnegie. Because versions of Rockefeller interest theory appear throughout the alternative medicine literature, sometimes in the form of almost Trilateralist conspiracy theories, this version of interest theory warrants some analysis even if most of it will be rejected. Consider three central propositions of this theory: 1) industrialists such as Rockefeller and Carnegie wanted to have healthy workforces and to deflect attention from environmental and social aspects of health, so they supported research on biomedicine; 2) their foundations completely reshaped the contours of medicine, particularly in the United States; 3) by the middle decades of the twentieth century, the Rockefeller financial empire had become interwoven with the chemical and pharmaceutical industry, and as a result Rockefeller financial interests—via its foundations and other ties—pushed cancer research toward chemotherapy.

Let us consider the propositions as research hypotheses. First, to limit the argument to the United States, what were the motivations of John D. Rockefeller, Sr., in allowing large amounts of money to be directed through foundations into medical research? The historical record hardly depicts the man as a mastermind who manipulated the medical profession as his puppet. The son of a "quack" doctor, Rockefeller was a believer in homeopathy who had a homeopath as his personal physician. He even tried to get the University of Chicago to include homeopathy in the curriculum in his offer in 1894 to fund a medical school for the university. According to historian Andrea Blumenthal, "This alliance is thought not to have materialized because the senior Rockefeller, devoted to homeopathy, did not want to support an allopathic school, and also because [Rockefeller foundation director Frederick] Gates was skeptical about how much control the Rockefeller philanthropies would have over the research done by an independent university" (1991: 69). The tendency to speak in terms of "scientific medicine" rather than homeopathy or allopathy emerged in part as a way of getting Rockefeller, Sr., to go along with the spending plans advocated by the medical reformers and their allies. The latter included John D. Rockefeller, Jr., and Frederick T. Gates, the former Baptist minister who became the captain of Rockefeller philanthropy.

Rockefeller, Sr., had little to do with the direction of medical funding decisions, and similar picture emerges for Andrew Carnegie. As historian Howard Berliner summarizes, "The actions that foundations took were more heavily influenced by the staff of the foundation than by the founders. Frederick T. Gates (not John D. Rockefeller, Sr.) and Henry S. Pritchett (not Andrew Carnegie) were actually responsible for the actions that were taken in the founders' names" (1985: 5). Why did Rockefeller, Sr., go along with the recommendations of Gates and his son? According to medical historian Harris Coulter, the homeopathic profession at the time was itself divided among reformers who advocated a more scientific approach to medicine and traditionalists who advocated the orthodox Hahnemannian line. Coulter adds, "According to one undoubtedly reliable report, [Rockefeller, Sr.] was dissatisfied at the inability of the homeopathic institutions to teach and promulgate the Hahnemannian doctrines."[4]

Rockefeller, Sr.'s, support for scientific medicine was therefore motivated by considerations other than those suggested by the first proposition. As for Carnegie, it is well known that his Scottish Calvinism became translated into the gospel of wealth, which included support for medical education. Rockefeller endorsed Carnegie's gospel of wealth, and according to Berliner he was "deeply religious" and "did not want to die disgraced" (1985: 13, 18). Although the effects of the foundation spending might have been to legitimate capitalism, transform robber-baron wealth into symbolic capital, improve the health of the working class, and so on, the motivation may have been simpler.

However, the foundation directors had broader goals in mind when they channeled money into medical research, and some of these goals were linked to a capitalist vision of the world. Gates wrote of disease as the "supreme ill of human life" and the "main source of almost all other human ills, poverty, crime, ignorance, vice, inefficiency, hereditary taint, and many other evils" (Brown 1979: 128). As Richard Brown argues in *The Rockefeller Medicine Men,* "In China, as throughout the world, the Rockefeller philanthropists soon concluded that medicine and public health by themselves were far more effective than either missionaries or armies in pursuing the same ends" (124). One consequence of their medical evangelism, perhaps not entirely intended by the philanthropists, was that the medicalization of poverty and social issues also contributed to their depoliticization. Brown argues that Rockefeller Institute money systematically precluded research on the relationship between social or environmental factors and disease. Of the 650 researchers associated with the Institute during the early decades of the twentieth century, very few understood "the role of society and environment as forces affecting the very diseases they studied" (129). As medicine became rationalized, the new scientific medicine also became more restricted and reductionist by focusing the understanding of disease on biological factors rather than seeing it as a mixed phenomenon that involved a social and environmental etiology as well. However, as historian Paul Starr points out, many socialists supported the new scientific medicine, and likewise capitalists such as corporate managers were very concerned with public hygiene (1982: 228). Consequently, concern with public health, which was salient in the Rockefeller philanthropy during the years prior to World War I, could provide a bridge across political and class barriers.

In short, there is no compelling evidence to support the view that Rockefeller, Sr., or Carnegie supported foundation spending for scientific medicine in order to serve their industrial interests in a narrow way, such as having a healthier working class. Although the foundation spending may have had positive consequences for their industrial profits, and it may have legitimated a capitalist vision of the world, the direction of the spending was not guided by these interests. Furthermore, there was no reason why the foundation money would not support research on the infectious theory of cancer. Gates had read *Principles and Practices of Medicine,* by Johns Hopkins professor William Osler, and he had become fascinated by the germ theory of disease (Berliner 1985: 56). As industrial companies now know, research on chemical carcinogens might lead to studies of workplace hazards that could cut into corporate profits, whereas research on infectious origins of cancer would not lead to this kind of question.

In fact, one consistent institutional home for research on tumor virology in the United States during the first decades of the twentieth century was the Rockefeller Institute for Medical Research, where not only Rous but other major virologists did their work. Furthermore, Rous occupied a privileged place as a close friend of Simon Flexner, the director of the Rockefeller Institute and a protégé of the Johns Hopkins medical school dean and Rockefeller insider William Welch. Flexner was close enough to Rous that he chose him in an unsuccessful attempt to make the virologist his successor as director of the Institute.[5] Flexner also sponsored cancer research on a variety of potential causal factors such as heredity, and he even supported a study on diet and cancer in mice, which, although not successful, showed his willingness to pursue new ideas (Blumenthal 1991: 262–63). Thus, the Rockefeller researchers were exploring a number of approaches to cancer, and tumor virology was one of them. Although the bacterial theories were not explored in the Rockefeller Institute, the Rockefeller family was on friendly terms with William Coley.[6] In short, there is no evidence that Rockefeller funding or interests led to the rejection of infectious theories of cancer.

Similar problems emerge for the second of the three propositions, that the foundation spending completely reshaped the contours of American medicine (and to some extent medicine in other countries as well). The debate over the relative influence of foundation money on the rise of scientific medicine in the United States has to some extent centered on the influence of the report on medical education prepared for the Carnegie Foundation by the educator Abraham Flexner, the brother of Simon Flexner (Flexner 1910). During the first decades of the twentieth century, medical reformers advocated much higher standards to bring medical education in the United States up to the levels achieved in Britain, France, and especially Germany. The Flexner report was a manifesto of this movement for reformers, whose work included centralizing the American Medical Association, increasing AMA control over state boards, tightening licensing requirements, pushing through increased science requirements in the medical education, and instituting a ranking system for medical schools (Brown 1979). Many medical schools merged or closed because their fee-based systems were not enough to cover the costs of the expensive laboratories that the new scientific medicine required. By the 1920s there were half as many medical schools as there were at the turn of the century. Financial support for the new medicine poured in from foundations, especially those of Carnegie and Rockefeller. The standardization of medicine meant that minority perspectives were often excluded. Although a few black universities were anointed to continue to produce a small elite of African-American doctors, many others were closed down and the doctor-patient ratio for African Americans fell considerably. The Flexner report also recommended closing the three women's medical colleges. A number of alternative medical traditions—such as homeopathy, naturopathy, and midwifery—were also driven underground or into highly marginalized positions.

Although many of the recommendations of the Flexner report were carried out, it is important to keep straight the lines of influence. The Flexner report was not a vector of transmission from industrial interests through the foundations to the medical profession. Instead, the direction of influence ran more the other way: the American Medical Association was involved from the beginning in the planning and eventual contours of the report, even to the point of prevailing over Abraham Flexner when he recommended regional norms and the medical organization wanted national norms (Berliner 1985: 105–8). In fact, several years prior to the researching of the Flexner report, the AMA had already formed a Council on Medical Education that called for many of the reforms that Flexner's report later advocated (Coulter 1987: 444–45). For these and other reasons, a growing consensus in the history of medicine holds that the impact of the Flexner report has been overrated. Historian Donald Fleming summarizes this view:

> He [Abraham Flexner] was eloquently articulating, with cautionary examples to the contrary, an ideal that was already on the way to prevailing if he had never published a word on the subject. The independently potent example of Johns Hopkins had to precede his own efforts, as he would have been the first to insist. Nevertheless, Flexner strengthened the hand of those who were fighting to upgrade their own medical schools and badly in need of reenforcement from an aroused public opinion. And if the scandalously bad schools were already falling by the wayside, he undoubtedly fostered mass suicide among the survivors. (1987: 208–9)

This passage points to the complexities of the relationships between the medical profession and the foundation initiatives. In general, the historical record suggests that if capturing occurred, it was in the form of the influence that medical leaders such as Simon Flexner and William Welch had over the funding tastes of Rockefeller, Jr., and Gates, who in turn influenced the largesse of Rockefeller, Sr.

There is some evidence to support the shaping role of the foundations in one aspect of the transformation of American medicine: the transition to the full-time plan. Although the topic was not part of the original Flexner report, Abraham Flexner performed a subsequent study at the request of Gates, who became a supporter of the idea after hearing a lecture from another medical reformer. The full-time plan would put medical professors on standard salaries that would replace private-practice income. From Gates's viewpoint, this change was crucial if professors were to be freed from clinical practice so they could devote their time to teaching and research, and therefore to complete the transition to scientific medicine (Berliner 1985: 142). However, the AMA and numerous other quarters of the medical establishment resisted the full-time plan vociferously. For example, the Harvard doctors successfully fought the Rockefeller foundation's attempt to put them on strict standard salaries with no outside income from private practice. The alternative developed at Harvard eventually became the model for the modified full-time plan that became dominant in the United States.[7] However, even though the full-time plan represents a strong case of foundation intervention in the development of the medical profession, the reforms were carried out with the blessing and support of prominent members of the medical profession, with William Welch at the forefront.

Rather than viewing the foundations as having a determining role in the shaping of medical reform in the United States, historians have demonstrated the crucial role that medical reformers played in influencing the funding tastes of the foundation directors. Starr argues that the medical profession was undergoing a series of major structural changes that probably would have occurred without the support of the foundations. These changes include strengthened professional organization and hospital power, control of labor markets via licensing boards, privately endowed and publicly supported medical research, elimination of countervailing power in medical care by private and public organizations, and the emergence of specialties (1982: 229–31). Although foundation money certainly helped the process, according to Starr the rationalization of medicine in the United States was a relatively independent process rather than one directed by industrialists through their foundations. This position is supported by historian Robert Kohler, who has confirmed and delimited the general importance of foundation money and the Flexner report in the shaping of scientific research in the United States. As he comments, "Flexner and the AMA leaders did not originate reform; they only routinized it" (1982: 126; also 1991). In short, the foundations provided a spur to the rise of scientific medicine in the United States, but they did not cause it.

Consider now the third argument, that as the Rockefeller empire developed financial interests in the chemical and pharmaceutical industries, from behind the scenes it pushed cancer research toward chemotherapy. Note that the place and time is different for this argument: the question of chemotherapy takes us to the Memorial Hospital during the 1930s rather than the Rockefeller Institute and the medical schools during prior decades. There is some evidence that by the 1930s the Rockefeller empire was building ties with huge European pharmaceutical empires, and that these organizations opposed the infectious theory of cancer. In Germany the microbial therapy of Günther Enderlein was, according to the Swedish doctor Erik Enby, "viewed by Nazi doctors as direct competition for the chemotherapy industry" (1990: 8). This comment suggests a connection between the German chemical and pharmaceutical industry and the rejection of Enderlein's microbial approach to chronic disease, although through the Nazi party rather than foundations as in the United States. Enby does not provide more details or references, but there is some historical documentation in support of the complicity of the great chemical cartel I. G. Farben (whose heirs today include Bayer, BASF, Hoechst, Agfa, Cassella, and Kalle) in the Nazi build-up of power or the Nazi regime.[8]

Ralph Moss suggests one piece of historical evidence for a linkage between I. G. Farben interests and Rockefeller money in the following passage:

> It is interesting to note that in 1926, one year before the Rockefellers began their systematic contributions to Memorial, Frank Howard, a vice president of Esso, paid his first visit to the I. G. Farben laboratories. He later said that he was "plunged into a world of research and development on a gigantic scale such as I had never seen" (cited in Borkin 1978). He soon discovered that the Germans were already deeply involved in cancer research. (1989: 393)

Moss cites a report from Howard that acknowledges Germany's leadership in the drug industry of the time and its advances in cancer research. He comments, "In 1927 the Rockefellers greatly expanded their interest in pharmaceuticals when Standard Oil of New Jersey (Esso), which was dominated by the Rockefeller family, signed an extensive cartel agreement with the German I. G. Farben Company" (1989: 392).

Although the two great globalized sociotechnical systems did develop a relationship during this period, it seemed to involve mutual outmaneuvering and rivalry as much as hidden conspiracy. As Joseph Borkin writes in *The Crime and Punishment of I. G. Farben,* the idea to build a cartel from the various German chemical companies came to Carl Duisberg of Bayer during a visit to the United States in 1903 to lay the cornerstone for a new Bayer factory in Rensselaer, New York. Duisberg learned about the Standard Oil trust during this trip, and he brought back with him the model of Standard Oil for trust-building in Germany. During the 1920s Standard Oil became interested in I. G. Farben's project to produce synthetic oil from coal, and I. G.—worried about its finances and aware of the technical problems in the research—sold its world rights to Standard Oil. Standard Oil also appeared to be the dupe of I. G. Farben during their various swaps of the 1930s, when Standard Oil sold I. G. a gasoline additive "not knowing" that it was going to be used for the Luftwaffe. However, Standard Oil apparently evened the score when toward the end of the war it persuaded I. G. Farben to transfer patents to it in order to avoid confiscation by Allied governments. Those patents remained in Standard Oil hands until the U.S. courts ruled in 1948 that the American government had the right to seize the patents. The government did seize the patents, and then when it auctioned them off, Standard Oil bought them back.

Thus, the historical research to date suggests that Rockefeller/I. G. Farben connections are ambiguous and do not provide a strong basis for the argument that the Rockefeller family had substantial pharmaceutical interests via the German cartel. It is possible, however, that Rockefeller interests in the pharmaceutical industry came through another channel, the pharmaceutical behemoth Burroughs-Wellcome. In *Dirty Medicine,* medical journalist Martin Walker writes,

> Over the last half century, medical education and research in Britain have been dominated by the interests of Wellcome and Rockefeller. . . . Following the recommendations of Flexner's European report, Rockefeller money began to pour into British medical research and education, reshaping the foundation of medicine in Britain as it had in America. (1993: 215)

And yet, the connection between Rockefeller and Burroughs-Wellcome money is tenuous. Walker points to the fact that Henry Wellcome's will of

1932 appointed John Foster Dulles and Allen Dulles, then of the Rockefeller-affiliated law firm of Sullivan and Cromwell, to handle legal matters relating to the will and the continuing financial empire (217). He also argues that the persons associated with the pharmaceutical company were involved in the Trilateral Commission (221). However, these are the extent of the apparent linkages.

In both the I. G. Farben and Burroughs-Wellcome cases, two problems remain. The nature of the connection between the Rockefeller and the other corporate empires remains weak. In other words, the researchers do not point to direct investment in specific pharmaceutical companies that in turn had a financial interest in cancer research. Furthermore, even if one were to find that the Rockefeller financial empire had developed a substantial financial interest in pharmaceutical companies by the late 1920s or early 1930s, there would still be several steps to make before one could claim that these financial interests had any impact on cancer research. Moss cautions that although Rockefeller funding of the Memorial Hospital began in the 1920s, the Rockefellers continued to support Ewing and there was no immediate change in leadership that can be attributed to Rockefeller influence (1989: 391–92). However, in the 1930s Howard joined the Memorial Hospital board, and Rockefeller influence may have grown after that point (393).

The question emerges, then, did persons associated with the Rockefeller family fortune play any role in the shift of cancer research toward chemotherapy? Documents reveal that in 1936 the General Education Board gave three million dollars to Memorial Hospital for a building, and in 1939 Cornelius Rhoads, who had worked in Simon Flexner's laboratory at the Rockefeller Institute, became the director of the hospital.[9] In 1941 the foundation gave an additional $120,000 to the hospital, with the reason that it had improved with the appointment of Rhoads, and from the 1940s on there is substantial interaction between the Rockefeller foundation and Rhoads.[10] However, the money trail does not prove anything; more interesting would be information on Frank Howard's motivations and his role (if any) in Memorial's transition to chemotherapy. Furthermore, the interaction between Rhoads and the Rockefeller foundation was not always favorable to Rhoads. In 1949 he wrote to Abraham Flexner a memo that proposed changing the staff organization of the hospital to a full-time system to make it a modern research-and-teaching institution, and Flexner passed on the memo to Rockefeller, Jr., with his enthusiastic endorsement.[11] The foundation apparently declined the offer. Nevertheless, this failed venture may turn out to be a minor footnote compared to the big picture: Rhoads pursued the new paradigm of chemotherapy with unwavering faith. This work entailed close ties with and praise of the pharmaceutical industry, including a contract with Burroughs-Wellcome to synthesize antimetabolites.[12] But this is as far as the rather confusing trail goes at this point.

To summarize the arguments regarding this version of interest theory, the first two propositions did not fare well when compared with the historical record. (Those propositions were that Rockefeller and Carnegie supported research on biomedicine in order to have healthy workforces and to deflect attention from social and environmental aspects of health, and that their foundations achieved those goals by reshaping the contours of medicine.) A reasonable argument can be made that those two propositions have been falsified. However, there is still no solution to the third proposition. Did the Rockefeller empire develop clear financial interests in the nascent pharmaceutical industry, and from behind the scenes did these interested parties push cancer research toward chemotherapy through involvement in organizations such as Memorial Hospital? It is possible that future research will reveal the answers, but it is also possible that we are faced with the situation of "queima do archivo" (burning the archive), to use the colorful Brazilian phrase. My conclusion, then, is that notwithstanding the false starts of the first two propositions, the third proposition leads to mixed results, including some positive results such as the discovery of new historical facts and the provision of a framework in which to locate them. Still, this version of interest theory does not provide an answer to the question of why the infectious theory of cancer faded into such an unorthodox position. Theoretically, it would have been possible to link an infectious theory of cancer to the interests of the pharmaceutical industry via antibiotics. To solve the puzzle, one needs additional explanatory resources beyond a version of interest theory that is limited to the shaping role of Rockefeller and other outside industrial interests.

Interest Theory, Version II

The second version of interest theory focuses more on the development of economic interests internal to the emerging profession of cancer treatment. This much more complicated approach is developed in Ralph Moss's influential book *The Cancer Industry.* There is no question that professional interests emerged in the late nineteenth century around surgery as the primary treatment modality for cancer. Surgeons who earned a substantial amount of their income from cancer clearly had a financial interest in the continued existence of their livelihood and practice. They would be highly motivated to question a theory and treatment (such as bacteria and vaccines) that could conceivably put them out of business. However, there was room for alternative therapies when surgery was not possible, and radiotherapy came to occupy this field rather than bacterial vaccines and other immunotherapies. The age of the therapies was not a factor: whereas there was a long research tradition on microorganisms and cancer, X-rays and radium were discovered during the 1890s. Nor was safety a factor: by 1910 the carcinogenic potential of X-rays was already known, and by the 1920s the dangers of radium were known.[13] How is it that radiotherapy carried the day when researchers such as Coley, who had introduced X-ray therapy to Memorial Hospital, were pointing the way toward much safer and potentially more efficacious immunotherapies?

Following the Memorial Hospital's official historian Robert Considine (1959), Moss argues that interest in radium therapy at the hospital developed dramatically after industrialist James Douglas gave a large gift in 1913 and attached several strings to the gift (1989: 65–66). According to Considine, "Douglas insisted first that his personal friend and physician, Dr. James Ewing, be made chief pathologist (later medical director) of the hospital; second, that the hospital treat only cancer patients; and third, that it routinely use radium in that treatment."[14] Ewing returned the favor by dedicating his book *Neoplastic Diseases,* in which he sounded the death knell for microbial theories of cancer, to his friend and benefactor.

Because Douglas was president of the Phelps-Dodge copper-mining empire and because he became involved in the mining and processing of radium, one might interpret his intervention as an example of the financial interests of the mining industry shaping the course of cancer research. However, there is as yet no evidence that the Dodge-Phelps empire benefited from diversification into radium. Instead, radium mining became an industry of its own, with its own beneficiaries and interested parties, including Douglas and the nascent cancer industry.

Douglas's sincere belief in the medical efficacy of radium, even to the point of drinking radium-treated fluids that may have led to his death in 1918, no doubt facilitated the creation of a radium industry. His daughter Naomi had developed breast cancer in 1907, and after five unsuccessful operations they decided to try radium in 1909. That attempt, too, was a failure, but after her death in 1910 Douglas continued to work to make radium available as a treatment for cancer. Nevertheless, in the great American tradition of combining evangelical zeal with the profit motive, Douglas and his allies in the medical profession developed a large financial stake in the business. In 1913 he cofounded, with the U.S. Bureau of Mines, the National Radium Institute. His partnership included Howard Kelly, who—along with William Halsted, William Osler, and William Welch—was known as one of the four founding fathers of the Johns Hopkins Medical School. According to the radiologist and historian Juan Del Regato,

> Douglas and Kelly bought twenty-seven claims in Paradox Valley, Colorado, and the Bureau of Mines undertook studies of ores, locations, methods of extraction, and production. Three years later, each partner received 3.75 grams of radium, which were eventually utilized at the Memorial Hospital of New York, the Huntington Memorial Hospital of Boston, and the Johns Hopkins Hospital of Baltimore. (1993: 69)

The value of a gram of a radium was $100,000 to $150,000 in the currency of the time. Kelly built a private hospital in Baltimore, gave radium treatments to Johns Hopkins cancer patients, and according to historian Bertram Bernheim, "became the owner of more radium than any other doctor in the nation" (1948: 23). In 1917 the first report of radium therapy at the Memorial Hospital reviewed 424 cases and claimed that it had resulted in at least 120 cases of complete regression at some point in the course of the disease (Janeway 1917: 222). By the late 1920s the Memorial Hospital, which became known as the "radium hospital," had eight grams of radium as well as the original radium produced by Marie Curie, who had to travel to America to seek financial help to purchase additional radium for her research.[15]

By 1924 the hospital's radium department became the single greatest source of income (Moss 1989: 67). Although the harmful side effects of radium were known by the 1920s, it continued to be used.[16] Studer and Chubin note that when Congress made its first appropriation for the NCI in the late 1930s, it stipulated that half the funds would go to the purchase of radium. Examination of Congressional testimony suggests the appropriation for radium was probably due to the suggestion of cancer researchers, not from members of Congress from states with relevant mineral deposits.[17] In other words, by the 1930s the therapy had stabilized and developed an inertia of its own.

Large corporations also developed an interest in the continued use of radiotherapy. According to Rhoads, General Electric was largely responsible for developing the X-ray for diagnosis and treatment.[18] As Moss comments, "Once millions of dollars are invested in capital equipment, there is strong inducement to use that equipment, despite newer information suggesting its use should be curtailed" (1989: 67). Radium remained a key component in orthodox cancer treatment long after Ewing, Douglas, Kelly, and their partners were dead, and it was only abandoned when radiotherapists found less dangerous radioactive substitutes that did not threaten the profession of radiation oncology.

Nevertheless, by the 1940s Memorial and other hospitals were taking a new direction that would ultimately provide an alternative to radiotherapy for millions of cancer patients. In 1939 Cornelius Rhoads left the Rockefeller Institute to become the director of the Memorial Hospital, where he led the transition to chemotherapy. During World War II he served as the chief of research for the Chemical Warfare Service, and the secret military trials with nitrogen mustard on 160 patients during the war contributed to his knowledge and advocacy of chemotherapy (Moss 1989: 393–94). During the period of 1941 to 1945, he also served as the American Cancer Society's director, which contributed to his leadership role and general influence in the field. He resigned that position after questions were raised about the appropriateness of his directing one organization that gave money to another for which he had a vested interest, namely Memorial Hospital.[19]

As a leader of the movement to develop chemotherapy in postwar America, Rhoads's directorship of the Memorial Hospital and his subsequent founding of the Sloan-Kettering Institute represented a general transition in research programs on cancer therapy. Rhoads, of course, was not a lone participant in this transition; other prominent cancer researchers such as Sidney Farber also played a leadership role (Moss 1995: 19). But Rhoads had a flair for stimulating public attention and corporate sponsorship. By 1949 he was on the cover of *Time Magazine,* and he was deploying new metaphors to stir up public enthusiasm for chemotherapy much as it had been stirred up earlier for radium.[20] Cancer was like a weed, he argued, and the older therapies of surgery or radiotherapy treated the weed by cutting it or burning it. Clearly, he viewed these therapies as relatively crude and empirical, because they did not attack the root cause of cancer. The alternative that he defended was to control the "biochemical soil" in which cancers arise. He drew a parallel between cancer cells and bacterial pathogens; he hoped that chemotherapy would selectively destroy cancer cells the way antibiotics destroy bacteria.[21] Another of Rhoads's metaphors gives a better picture of his approach; he described chemotherapy as "Waging the Chemical-Biological Cancer War," a metaphor which is not surprising in light of his wartime experience.[22]

Although Rhoads compared the chemotherapeutic treatment of cancer cells to the antibiotic treatment of bacterial pathogens, this was about the extent of his use for infectious theories on the topic of cancer. His hostility to bacterial theories and therapies, both those of Coley and Livingston, was demonstrated in the previous chapter. Although his archived papers are almost silent on the topic, one exception is a paper given in 1950 before the Sterling-Winthrop Research Institute in Rensselaer, New York, where Rhoads rejected bacterial theories outright.[23] He also remained skeptical that viruses caused any cancers in mammals. He did support a conference on oncoviruses and he recognized the work of tumor virologists such as Peyton Rous, but he thought the viral research program for cancer was a failure. He argued that the research program had not generated successful cures for cancer, and he suspected that purported oncoviruses might have been confused with other cell-free filtrates.[24] Instead, his imagination was captured mainly by the possibility that viruses could be developed as possible cytotoxic agents similar to antibiotics or phage, and he wrote about possible "oncolytic" viruses that could be directed only at cancer cells.[25] In other words, chemotherapy guided his thinking to such an extent that he viewed the potential of viruses in similar terms as selective cytotoxic agents.

It follows that the chemotherapeutic research program would lead to cozy relationships with the pharmaceutical industry. These relationships served as keystones in what Moss aptly has termed the cancer industry. In a talk on "Industrial Science and Cancer Research," given for the General Electric Science Forum and broadcast from the "electric city" (Schenectady) in 1947, Rhoads praised industrial laboratories for turning sulfa drugs into reality, and he credited the pharmaceutical and chemical industries for contributing to health.[26] In another paper he praised the pharmaceutical industry for contributing to three key areas of cancer research: steroids, nitrogen mustard, and antimetabolites.[27] The work with Burroughs-Wellcome has already been mentioned, and in general Memorial Hospital became a center of drug testing for the new chemotherapy (Bud 1978: 442ff.; Moss 1989: 83ff.).

However, the pharmaceutical industry was not the only industry with which Rhoads developed relationships. He also sought out benefactors for his research program from industrialists outside the pharmaceutical industry. At the suggestion of Frank Howard—the vice president of Esso and Memorial Hospital board member mentioned above—Rhoads met with Charles Kettering of General Motors. Negotiations with Kettering and William Sloan eventually led to the founding of the Sloan-Kettering Institute for Cancer Research in 1945, of which Rhoads became director.[28] In 1951 he developed "A Plan for Corporate Participation in the Memorial Cancer Program," in which he sought donations of $10,000 per year from a hundred corporations in return for services that would review two types of data: cancer cases of individual executives and cases of alleged corporate responsibility for cancer problems.[29]

As closer relations with the corporate world developed, Memorial Sloan-Kettering and other cancer research institutions adopted organizational models from industrial research. Rhoads vehemently denied charges that the Sloan-Kettering donation implied that he was setting up an industrial research laboratory, and he argued in his defense that the research produced at Sloan-Kettering would be published and shared freely, unlike the proprietary research of industrial laboratories.[30] However, the models of industrial research laboratories influenced cancer research in other ways. In an examination of Sloan-Kettering and the Philadelphia-based Institute for Cancer Research, science studies analyst R. F. Bud (1978) demonstrated that during the 1940s these two leading institutions imported organizational models from industrial research laboratories. He argues that this organizational style favored the emphasis on chemotherapy that emerged during the period, when "teams with primary responsibility for chemotherapy became the largest units" at both Sloan-Kettering and the Philadelphia-based Institute for Cancer Research (1978: 441). The importation of industrial research culture—with its values of standardization, hierarchical and centralized organizations, and synthetic products—cloaked the transition toward chemotherapy in a sense of legitimacy because it seemed "naturally" to be the best way to organize work and thought. Although other lines of research were allowed to coexist, they were given relatively low priority.[31]

Rhoads and Memorial Sloan-Kettering may represent an extreme case, but the transition to widespread testing of chemotherapies, the appeal to corporate funders for support, and the industrial organization of research was widespread. The legacy for cancer research today is significant interlock among major corporations, the major cancer research organizations, and the government agencies that fund and regulate cancer research and therapies. In an analysis of the board membership of the Memorial Sloan-Kettering Cancer Center, Moss concluded that the pharmaceutical industry "has great influence on the MSKCC board, especially on the select Institutional Policy Committee" (1989: 443). He showed that in 1979 "seven out of nine—or 78 percent—of the members of the Institutional Policy Committee were affiliated (or interlocked) with companies with a direct interest in the cancer drug (or diagnostics) market" (444). Likewise, Moss notes that drug companies such as Bristol-Myers Squibb directly influence cancer research by giving awards and grants, supporting lectures, updating cancer textbooks, and supporting clinical studies of their patented agents (1995: 79).

To some extent, the current situation of interlocking corporate, non-profit, and government boards, grants, and research projects makes the Rockefeller version of interest theory irrelevant. Rather than examine the tenuous proposition that industrial interests shaped the history of cancer treatments and research, I believe the more pertinent questions concern the nature, structure, and interests of cancer research and treatment as it developed into an industry unto itself. The cancer therapeutics market, which itself is only a portion of the entire amount of money spent on cancer prevention and treatment, is growing at a rate of about 12 percent per year and is currently valued at about ten billion dollars per year (Moss 1995: 75). Cancer therapeutics therefore represent a significant sector of the pharmaceutical industry.

Thus, the second version of interests theory would hold that over time the cancer research community developed its own internal cultural logic that was interwoven with its own financial, institutional, and professional interests. The cultural logic included a core of research values such as extreme skepticism toward the view that viruses play anything more than a minor role in human and mammalian cancers, and that other microorganisms, particularly bacteria, are anything other than secondary infections. The consensus view can be justified by pointing to a long history of cancer research that dates back to the founding moments at the beginning of the twentieth century, but this view also ignores another research tradition that became marginalized and that can also be justified by its own body of evidence. Clearly, the marginalized view has quantitatively less evidence at this point, but as I will show in the next chapter, it still points to some anomalies that suggest bacteria and perhaps fungi play an overlooked contributing role in cancer etiology.

Science studies of specialty formation suggest that there is a general process in which autonomy increases as a field of research becomes more defined, routinized, and driven by a generally accepted research program.[32] The case of the cancer research community confirms this pattern, but I would add that there is a formative period when the basic directions of the research program are set in place (such as the refusal to see cancer as a metabolic, nutritional, or infectious disease and one that could be treated by vaccines, sera, and nutritional therapies). Furthermore, as the field of research became increasingly technical and specialized, the choices that were so evident at the beginning were largely forgotten. The controversies that Ewing and Coley were part of became lost to the historical unconsciousness as the noninfectious nature of cancer became common sense to members of the culture. Yet, the consensus did not emerge from entirely internal, intellectual processes such as the consideration of evidence. Rather, the commonsense consensus was also adaptive to the political/economic ecology that provided incentives for therapies oriented toward X-ray machines, radium, pharmaceuticals, and other industrial products.

Although apparently guided by and justifiable in terms of evidentiary criteria, the general contours of orthodoxy and heterodoxy remain products of a mix of evidential considerations and interested history. Many actors were attempting to build their networks, but those who advocated radium and chemotherapy prevailed over those who advocated bacterial vaccines, not because the former were better at interesting allies or marketing their therapies, nor because the theoretical choice had been settled unambiguously by evidential considerations, but because they offered therapeutic research programs that were better adapted to the ecological pressures of the industrial-capitalist society in which they were operating. Thus, in contrast to theories of science such as actor-network theory (Latour 1987, 1988), I suggest that although the new networks of official cancer therapy and research contributed to the reproduction of an industrial-capitalist order, they did not make that order. Rather, that order exerted a shaping ecological pressure on the contours of the research culture. In short, the seamless web of content/context has to hang itself somewhere.

The concept of a research culture renders more complex and sophisticated the analysis of interests in the history of science, technology, and medicine. Like a language, a culture is capable of change, but its basic values (like a language's deep grammatical structures) tend to remain stable over time. The United States, for example, is characterized by a great degree of religious, ethnic, lifestyle, and other diversities. Yet, seventeenth-century British and Dutch Calvinism has cast an imprint that remains evident throughout the culture, and many of the basic values and institutions described in Alexis de Tocqueville's *Democracy in America* are still recognizable today. The comparative perspective of de Tocqueville is crucial for understanding these deeper structures of the culture. In a similar way, over time a research community develops a culture that can be traced back to its originary history and the historical conditions in which it was embedded during its early development. Over time the culture also develops a resilience and logic of its own, such that it responds to outside interests from the perspective of its values and logic.

Thus, my argument is that different sorts of interests—or, if one prefers, ecological factors—shape the development of a research culture, but not in a direct sort of way that reduces the responses of the members of the research community to interest dopes. It is much more like the relationship posited by Marshall Sahlins in *Culture and Practical Reason,* in which anthropologists developed a complex solution to the relationship between ecology and social organization. Rather than see changes in ecology as directly shaping aspects of the culture such as social organization, Sahlins defended a view similar to what I am articulating here: the changes in ecology were mediated through a culture that structured its responses via adaptive changes in social organization that tended to preserve the established cultural logic. In a similar way, financial interests do shape some research communities and research programs in powerful ways, but the interests become inflected as they reverberate in the historically developing cultures of research communities, disciplines, networks, and organizations.[33]

Monomorphism and Modernism

My analysis up to this point is still incomplete for several reasons. To begin, I have not yet explained why bacteriological researchers did not rush into the vacuum left by the cancer researchers. There is no reason why bacteriologists would not develop the research on ostensible viral and bacterial causes of cancer. One would presume that bacteriologists would have an interest in expanding their discipline to cover an important disease, but instead they generally ignored research on bacterial etiologies. To answer this question we need to turn to the history of bacteriology, or what is today known more broadly as microbiology, and the parallel history of the constitution of its culture with its attendant contours of orthodoxy and heterodoxy, or if one prefers, of commonsense consensus and jettisoned research programs. In the process, another layer to the question of the relationship between industry and cancer research will be explored: the embeddedness of the cancer research consensus in modernist cultural categories and values that were largely products of the industrial culture of the time.

What Glover may not have realized, but probably what Rife's partner Kendall knew all too well, was that by positioning the so-called cancer microbe as a pleomorphic bacterium that had viruslike or filterable stages, they had stepped right into the middle of a raging scientific controversy. To understand this controversy as it played out in the 1920s and 1930s, it is necessary to step back in time to the late nineteenth century and the originary moment of microbiology (then bacteriology) as a scientific research culture.

One of the heroes of some advocates of alternative cancer theories is Antoine Béchamp, considered to be the defeated and unsung nemesis of Louis Pasteur. Pasteur captured the imagination of his fellow citizens in his day, and he has continued to capture the imagination of historians and sociologists since then. At the whig history end of the spectrum, René Dubos's *Louis Pasteur: Free Lance of Science,* Pasteur is a saintly citizen whose masters are work and science. At the constructivist end of the spectrum, Bruno Latour's *The Pasteurization of France,* Pasteur's brilliance lies more in his ability to make himself indispensable to groups much more powerful than he. Public health, agriculture, and the food industry become interested in and then dependent on Pasteur's science and technology; he Pasteurizes France.

In the alternative medical literature there is a radically different view of Pasteur; he is a plagiarist and the nemesis of the crucified Antoine Béchamp.[34] Béchamp was a medical biochemist who at one point was a professor in Lille. He has all but disappeared from the more accessible and standard histories of microbiology, but this underdog of the history of microbiology retained a few loyal champions. For example, in an old book dating back to the 1920s, *Béchamp or Pasteur? A Lost Chapter in the History of Biology,* E. Douglas Hume argues that Béchamp deserves priority for many of the discoveries attributed to Pasteur. It is Béchamp, not Pasteur, who deserves priority for the experimental demonstration that fermentation is due to airborne living organisms, and therefore for the definitive refutation of spontaneous generation prior to Pasteur's conversion to the same view. It is Béchamp, not Pasteur, who discovered the cause of the silkworm blight of the 1860s. Only later did Pasteur change his theory and accept Béchamp's position. However, when Béchamp sought recognition for his discoveries, Pasteur and his network systematically excluded recognition of Béchamp's priority. In short, the narrative from some quarters of the alternative medicine community is that Pasteur and his network stole Béchamp's work, then rubbed him out from the historical record.

Why would a question of historical credit be important for science today? Because it points to an originary moment in the formation of the culture of modern microbiology, in which some ideas were so radically different that they were excluded from possibility. Rightly or wrongly (and I will argue something in between), the excluded ideas became pseudoscience.

A summary of many of Béchamp's key concepts can be found in *The Blood and Its Third Anatomical Element,* a book that stands out among his publications because it represents a mature formulation of his theory and because it is one of his few works that has been translated into English. The English translator—Montague R. Leverson, M.D., of the Baltimore Medical School—introduces Béchamp as "the Master" and Pasteur as "the arch plagiarist." Béchamp continues his campaign against Pasteur in the pages that follow. He also reveals a radical theory of microbiology that Pasteur and his followers never accepted, plagiarized, or wanted much to do with at all: the microzyma theory.

It is difficult to summarize the microzyma theory, partly because its language is so radically different from that of microbiology today. Reading Béchamp's work provides a good example of the problem of paradigm incommensurability or untranslatability of terms across radically different global theories. Nevertheless, the key assumptions of the microzymian theory can be approximated as follows:

1. The smallest unit of life is not the cell, but the microzyma, which is found in the cell cytoplasm and in the blood plasma. Microzymas can be seen in the microscope as little points, but according to Béchamp they are not, as critics have claimed, artifacts such as Brownian motion or bacterial spores. They can be found in the air, the earth, and in fossilized animal remains such as chalk.
2. Microzymas are specific to species and tissues; in other words, they are not all identical.
3. Microzymas survive the organism at death, and they contribute to the decay process.
4. Microzymas are capable of coalescing and causing the blood to clot.
5. When the overall state of health becomes compromised, microzymas can coalesce and form bacterial pathogens that cause disease. "Normal air contains neither preexisting germs nor the things which have been improperly termed microbes, supposed to ascend from age to age to parents resem-

> bling them" (Béchamp 1911: 393). Disease is caused by a change in the organism's overall state of health that leads to the transformation of microzymas into bacterial pathogens.

On his deathbed, Pasteur is said to have changed his mind about the infectious theory of disease and recanted with famous phrase, "The terrain is everything." Béchampians take the deathbed confession as proof of guilt and proof that their master was right. As I will argue in the next chapter, I think most of what Béchamp said was plain wrong. However, he articulated two ideas that in a somewhat different form have become accepted among the advocates of a bacterial theory of cancer: microbial pleomorphism is much more widespread than was first recognized; and latent microbial pathogens are capable of becoming pathological when environmental circumstances such as host nutritional status change.

Béchamp was an advocate of an extreme version of pleomorphism; he believed that bacteria are capable of changing almost limitlessly from one type to another. No one among the advocates of a bacterial theory of cancer accepted this position, although Rife came close in his discussion of eight major bacterial groups and Enderlein's protits recall Béchamp's microzymas. In Germany the contemporary counterpart of Béchamp among the extreme pleomorphists was Carl Nägeli. He believed that bacteria could not be divided into species because they undergo such profound changes of form over time and across environments. He also became involved in a controversy in Germany that was similar to that between Pasteur and Béchamp in France.

According to STS researcher Olga Amsterdamska (1987), botanist Ferdinand Cohn and bacteriologist Robert Koch formulated their position of bacterial monomorphism in opposition to the extreme pleomorphism advocated by Nägeli. The principle that is sometimes known as the Cohn-Koch dogma of monomorphism held that bacteria could be divided into specific species that maintained constant forms over time. Like Louis Pasteur, Koch is generally considered one of the founding fathers of bacteriology. Koch's studies of anthrax established the fact that spores isolated from pure cultures of the bacillus could cause disease, and in more general terms he showed that specific infectious microorganisms could cause specific diseases. Following the work of his mentor Jacob Henle, Koch supported the germ theory of disease, and he developed Koch's postulates for defining causation in disease.

Amsterdamska invokes professional interests to explain the controversy between monomorphists and pleomorphists. She draws on the research of historian Robert Kohler (1985) to argue that monomorphism was favored in part because of the need for quick identification and diagnosis in medicine. For example, Koch and his supporters advocated treating cholera by isolating and quarantining afflicted individuals based on laboratory analyses of bacteria. As Amsterdamska summarizes, "Monomorphism transformed laboratory observations of constant associations into causal generalizations, and thereby also assured their practical relevance" (1987: 666). In contrast, although the radical pleomorphist Nägeli accepted the germ theory of disease, he denied its medical relevance because he believed that bacteria could easily change forms. Nägeli was associated with the German hygienists, and they advocated treating cholera by improving sanitation.

Consistent with the theory of monomorphism was the secondary theory that bacteria reproduced by fission. It was not until 1946, when Joshua Lederberg and Edward Tatum published a paper in *Nature,* that biologists recognized genetic exchange as a means of reproduction in bacteria. In 1986 Lederberg published with STS analyst Harriet Zuckerman the paper "Postmature Scientific Discovery," which argued that in addition to the commonplace occurrence of scientists who make "premature" discoveries, it is also possible, as in the case of bacterial sex, for discoveries to be behind the times, that is, "postmature." One prime example (and perhaps not a surprising one, given Zuckerman's coauthor) was the case of the Lederberg and Tatum "discovery." Zuckerman and Lederberg argue that one of the major reasons for the delay of the discovery of bacterial recombination was the belief in monomorphism.

However, Amsterdamska (1987) demonstrated that the doctrine of extreme monomorphism was widely accepted only until the last decades of the nineteenth century. Thus, to account for the "postmature" discovery of Lederberg and Tatum one would have to find some other factors that did not include the doctrine of monomorphism. Amsterdamska's criticism is a significant one because her application of interests theory makes it possible to explain why monomorphism triumphed in the nineteenth century and why interest in bacterial variation reemerged at the turn of the century.

Why did bacteriologists become more interested in a modified, species-bound version of bacterial pleomorphism after the turn of the century? After the Koch program was more firmly established, bacteriologists began to tackle the problem of typhoid. In the first decade of the twentieth century, Rudolf Neisser demonstrated how the fermentation characteristics of a typhoid organism isolated from a patient were subject to change.

Amsterdamska argues that given the prevalence of typhoid and related diseases, and the difficulties associated with bacterial diagnosis, interest in bacterial variation then became much more widespread.

Even prior to 1900 Koch was not dogmatically opposed to the idea of limited bacterial variation. According to microbiologist Lida Mattman, in 1881 the Swedish researcher Ernst Almquist observed with Koch variations of typhoid bacteria (1993: 2–3). The German scientist apparently did not dismiss the observations, but, as Almquist later noted, Koch became too involved in his various research projects to follow up on them. Almquist did follow up on these and other observations, thus playing a founding role in the research tradition on bacterial variation. He argued that typhoid bacteria in food or water become an amorphous mass that regains its standard form and virulence once it enters the human host. He also observed evidence of what would today be called bacterial recombination—long before Lederberg and Tatum's "postmature" discovery.

Amsterdamska adds that there was some cultural variation in the controversy. Even during the nineteenth century, French researchers were relatively more open to recognizing bacterial variation than their German counterparts. She explains this difference as due to the French control of disease through attenuation in contrast with the "precise etiological determinations" of the Germans (1987: 670). Clearly, however, the French Pasteurians were not supporting the extreme pleomorphism of Béchamp, which, as with the Germans for Nägeli, was rejected.

To summarize, by the first decade of the twentieth century a consensus had emerged in bacteriology in favor of a modified theory of monomorphism that accepted limited bacterial variation for some species such as typhoid. This consensus continued more or less throughout the twentieth century, although there were differences of opinion regarding how much variation occurs, which bacterial species exhibit it, and the range of stages that could legitimately be counted as variation. The consensus that there are specific bacterial species was crucial for bacteriology to maintain its position as a key medical science that could contribute to the diagnosis, prevention, and treatment of disease.

Regarding the question of the bacterial theory of cancer, the timing of key research projects now becomes important. The studies of the late nineteenth century that supported the bacterial theory occurred when monomorphism was still very popular. Thus, to the extent that critics assumed monomorphism, they would have tended to see the contradictory bacterial cultures (cocci, rods, etc.) from cancer tissues as contaminants rather than different forms of the same microbe. Furthermore, the limited return to pleomorphism that became acceptable in the twentieth century still did not entail the extended, multistage models that Glover and Enderlein defended in the 1920s and 1930s. This became evident during a second major bacteriological controversy that erupted during the 1920s and 1930s and involved more or less this same fundamental issue: the extent to which microbes can vary.

In the second controversy, the new advocates of pleomorphism agreed that specific bacterial species exist and cannot be transformed into each other, and therefore they did not advocate returning to the radical position of Béchamp. However, they did advocate a strong form of pleomorphism within some bacterial species. This strong form of pleomorphism included two key but separable ideas: the observations of bacterial pleomorphism could be arranged into a life cycle (cyclogeny), and in at least some cases the cycle included a stage that was largely invisible under the light microscope and that could pass through a filter that blocked bacteria (filtrationism). During this period, which antedated the electron microscope, viruses were generally not visible through light microscopes, and filters were used to distinguish viruses from bacteria. As with prions today, there was a great deal of confusion about what viruses were: living agents or not, capable of surviving outside the cell or not, even real entities or merely artifacts. At that time, the term "filterable virus" was often used, and advocates of the theory that bacteria passed through well-defined life cycles like those of fungi were also known as "filtrationists." It was possible, however, to be a filtrationist without accepting the cyclogenic theory, although the two positions tended to coincide during this period.

In a second paper Amsterdamska (1991) takes up the cyclogenic/filtrationist controversy that began during the late 1920s and lasted into the 1930s. The controversy emerged over attempts to synthesize diverse observations into a general theory that bacterial variation occurs in life cycles similar to those of fungi. One of the key papers in the controversy in the United States was a review essay published in the *Journal of Infectious Diseases* in 1927 by Philip Hadley, an agricultural bacteriologist who after 1920 was at the University of Michigan Medical School. In addition to synthesizing previous work on bacterial life cycles, Hadley argued that morphological changes were linked to virulence, and therefore that the cyclogenic theory had tremendous implications for medicine. Furthermore, he supported the argument that the bacterial life cycle could, in at least some cases, include a viral phase. Another important figure—one cited in the literature and recognized in a historical comment by microbiologist Lida Mattman in her textbook on bacterial variation—was Rife's colleague Arthur Kendall, who, according to Mattman, was the founder of "a school of filtration" (1993: 209).

From the late 1920s into the 1930s, bacteriologists argued over the evidence for cyclogeny and filtrationism. In 1937 Hadley published a paper that summarized the results of both sides of the controversy and ceded some ground to the critics. However, the paper did not mark a closure of the controversy, nor did the controversy ever seem to achieve a decisive closure. "Rather," Amsterdamska argues, "as bacterial genetics replaced the studies of bacterial variation and dissociation, and as virologists came to a consensus that viruses are biologically completely different from bacteria, both sides to the life-cycle controversy became irrelevant and forgotten" (1991: 220). Mattman suggests that the filtrationist controversy was resolved by "the demonstration by [Emmy] Klieneberger and by [Louis] Dienes that filterable organisms could be grown on solid medium and their sequential reversion steps followed" (1993: 209). She concludes the chapter with the statement that all bacteria appear to be able to enter into a filterable stage, and it seems "strange" that the topic "was at one time a highly controversial subject" (214).

However, the fate of the cyclogenic theory was quite the opposite of the filtrationist theory.[35] A review essay published by the U.C. Berkeley biologist Werner Braun rejected the cyclogeny theory (1947). Likewise, in another review essay published a few years later, Klieneberger-Nobel summarily dismissed Enderlein's work: "Enderlein's book *Bakteriencyclogenie* presents a philosophical treatise rather than a discussion of scientific facts based on exact observations and is therefore not further discussed here" (1951: 93). Models of extended bacterial life cycles of the type proposed by Glover, Enderlein, and von Brehmer (a tradition continued today by Naessens) have been rare in the subsequent microbiological literature.

Why were the cyclogenist/filtrationist theories so controversial? Amsterdamska argues that Hadley's claims received attention because he made them medically relevant. His claims were disputed because even though he did not return to the radical pleomorphism of Béchamp or Nägeli, he questioned some of the basic assumptions of his contemporaries. He admitted that domesticated laboratory strains could be maintained in artificial conditions and that these strains were "amenable to consistent laboratory findings and have therefore become the favorite subject-matter of the systematists" (Hadley 1927: 289). However, he believed that the "free-living microorganism is potentially a kaleidoscopic thing, in which the power of responding to a changing environment by alterations in the body state . . . stands as its one most important attribute" (289). He suggested, "What we ought to wish to know about these cultures is, for instance, how the various types of pneumococci, of streptococci, meningococci, of B. coli, of the paratyphoids, of dysentery and other bacteria, have been (we might even say 'are being') formed" (1927: 290). Consequently, his colleagues should abandon "for the present at least, our vain attempts to perfect schemes of classification" (291). The bacteriologists' classificatory bible *Bergey's Manual,* which was first published in 1923 and was in its fifth edition by 1939, would have been thrown out the window.

Amsterdamska argues, "Hadley's claims affected in a direct manner the then-current work of members of the establishment in interwar bacteriology," including prominent bacteriologists at Johns Hopkins, Yale, and Harvard (1991: 209, 213). For example, Leo Rettger and Hazel Gillespie of Yale wrote,

> The questions at issue here are of extreme importance, and upon a satisfactory solution of them will depend the future development of bacteriology. If bacteria must pass through various life cycle phases, and if they manifest themselves in viable forms which are filterable, our conception of bacteriology will require a thorough revision, and with it the present conception of pathology of medicine. . . . [If not,] the threatened cataclysm in systematic bacteriology may be indefinitely deferred. (1933: 290)

Likewise, Martin Frobisher of Johns Hopkins wrote that systematic bacteriology based on the study of cultures subject to such variation "would seem based upon a fallacy," and "under these circumstances chaos exists in bacteriology."[36]

Similar comments frame the filtrationist debate between Kendall and his opponents that occurred at the 1932 annual meeting of the Association of American Physicians. Harvard bacteriologist Hans Zinsser attacked Kendall's filtrationist views and argued:

> If his surmise is correct, the entire structure of our attitude toward the biology of disease must be changed. It is almost as important as the theory of spontaneous generation, and nothing short of absolute proof should be accepted or we may risk making research even more difficult than it already is. . . . If his conclusions are correct, he has brought about a revolution in bacteriology, but we do not believe that he obtained the organisms from the filtrate after repeated invisible cultivation in the K medium. (Kendall et al. 1932, 68)

Zinsser could have been particularly affected by Kendall's research and his publications that claimed a high level of pleomorphism for the typhoid bacterium. In the early 1930s Zinsser was attempting to develop a vaccine for typhus, a disease caused by a similar bacterium, and if claims of extreme pleomorphism were extended to typhus, his efforts to develop a vaccine could have been complicated.[37]

Rockefeller Institute virologist Thomas Rivers had other reasons for questioning the filtrationist theory. In 1926 he announced the theory that viruses could not reproduce outside cells, and he added his own arguments and evidence against Kendall's filtrationist theory: "No one has brought convincing evidence that filterable virus exists in the absence of living tissue" (Kendall et al. 1932). Kendall apparently had few defenders, but ironically one was William Welch, who in addition to being one of the founders of Johns Hopkins and the president of the board of scientific directors of the Rockefeller Institute, was also the teacher of Peyton Rous.[38] In the exchange among Zinsser, Rivers, and Kendall, Welch stepped in with grandfatherly graciousness to note that however erroneous Kendall's interpretations were, he at least had contributed some new observations on the recognized phenomena of granules formed within bacteria (Kendall et al. 1932).

To summarize, cyclogenists like Hadley and filtrationists like Kendall threatened to destabilize the fundamental categories of bacteriology. But why did the bacteriological establishment place such a high value on having stable categories? One might propose a psychology of conservativism and inertia, or a quasisacred dogma of modified monomorphism, but a psychologizing explanation of this sort amounts to little more than sociological phlogiston theory. A better but still inadequate explanation is that the cyclogenist and filtrationist theories conflicted with the professional interests of medical bacteriologists, which rested on their role in fighting disease. This explanation is not completely adequate because, according to Hadley, patterns in bacterial variation could be correlated with medically relevant knowledge such as virulence. His theory could have enhanced the medical applicability of bacteriology. The problem seemed to be that he could not make the knowledge relevant enough so that it was worth accepting an overhaul of bacteriology and an effort to develop therapies that linked variable pathogenicity to microbial stages. Hadley had attacked a core idea of the bacteriological culture—a negative heuristic of its research program—the idea that bacteria could be safely classified into readily identifiable and relatively stable species and studied experimentally and systematically. Widespread and wide-ranging variation meant that the project of systematic bacteriology would have to be put on hold, and that efforts to develop therapies would be more complicated than at first anticipated.

More than a conflict of interests, the filtrationist/cyclogenist controversy involved a conflict of cultures. The heterodox position was at dramatic odds with the emerging culture of scientific medicine that included bacteriology. Across a wide variety of cultural domains—industrial production, medicine, literature, the arts, social theory, architecture, religion, education, social organization, politics, and so on—a modernist culture was forming in a pattern of mutual feedback and reinforcement. In a previous book, I emphasized closed system principles and equilibrium dynamics as characteristic of modernist culture (Hess 1995: ch. 3). In science, closed systems with equilibrium dynamics included the Bohr atom, Pavlovian conditioning, Parsonian functionalism, and bacterial systematics. Modernist culture also emphasized standardization (Martin 1994). During the modernist period of assembly-line production, Taylorist management, and functionalist architectures and social theories, standardization was positively associated with progress, not negatively linked to a cultural politics of conformity and hegemony. John D. Rockefeller's standardization of the oil industry and the medical profession's standardization of its education and licensing policies are only two examples of the value placed on standardization in the modernist period.

The bacterial research on cancer and its associated therapies conflicted with these often taken-for-granted values. The research failed to produce a standardized pathogen for cancer as in the case of other infectious diseases. Furthermore, it was linked to the filtrationist program, which threatened to destabilize the basic unit of bacteriological analysis. In a similar way the therapies that were associated with bacterial theories of cancer were not easily standardized. As Moss points out, in addition to the financial motive that favored the selection of radium and X-rays, radiotherapy was easier to prescribe in measurable dosages, whereas Coley's toxins were more difficult to administer (1989: 123). In the previous chapter, I reported on Helen Coley Nauts pointing to some of the difficulties in standardizing Coley's toxins and their application. With radiotherapies and chemotherapies, standard doses could be encoded into machines or syringes. Pleomorphic bacteria and variable vaccine potencies conflicted not only with the interests of assembly-line production of medical therapeutic technologies, but with a modernist culture of standardization.

In the United States the modernist culture of standard/scientific medicine was most advanced in elite, Eastern schools such as Johns Hopkins. These sites represented not only the starting point of the spending sprees of the Rockefeller and other foundations, but also the institutional home of some of the most outspoken critics of cyclogeny and filtrationism. Obviously the critics were not motivated by an unspoken allegiance to Rockefeller financial empire interests; instead, they had a standardized, modernist vision of science, nature, and society that overlapped with the industrial worldview and the emergent culture of scientific medicine.

There is also some evidence that the advocates of cyclogeny and filtrationism were either outsiders or located institutionally on the fringes of the establishment circles of the new scientific medicine. Historian Jonathan Harwood has documented an interesting intellectual divide among elite and nonelite geneticists in Germany, and he points to preliminary evidence for a similar division that may have existed in the United States between geneticists working in elite private universities and those working in the agricultural facilities of state universities:

> A casual prosopographical survey of these two groups suggests that those working in agricultural institutions were more likely to come from farm backgrounds, more likely to work in transmission genetics, and less likely to have "high cultural" interests than were those working in elite universities. (Harwood 1993: 356–57)

Likewise, in the case of the filtrationist/cyclogeny controversy it is suggestive that some of the advocates were located outside the elite, Eastern schools. It would be interesting to look more carefully at the institutions and backgrounds of the major defenders and debunkers of filtrationism and cyclogeny in terms of their proximity to the diffusing culture of scientific medicine.[39]

Whatever the internal social stratification dynamics of the cyclogenist/filtrationist controversy turn out to be, the controversy clarifies the meaning and position of the work of heterodox researchers such as Glover. When the Canadian doctor published research in the late 1920s that supported a microbial cancer agent with a high level of pleomorphism, a filterable stage, and an organization of stages into a life cycle—and when he added cyclogenist Felix Löhnis to his citations—he was stepping, perhaps naively, directly into a wider controversy. Furthermore, he was aligning himself with a side that was outside the main lines of the Eastern, elite establishment of bacteriology.

Other members of the filtrationist school probably were more aware of the linkages among pleomorphism, filtrationism, and the microbial theory of cancer. Ralph Mellon, a cyclogenist from the Highland Hospital of Rochester, New York, and his colleagues isolated pleomorphic tuberculosis microbes from tissues of Hodgkin's disease and sarcoidosis, a chronic disease similar to tuberculosis (Mellon and Fisher 1932; Beinhauer and Mellon 1938). Kendall, a key participant in the controversy, carefully appropriated the Rife research to defend his own position as a filtrationist without endorsing in print Rife's claims about a cancer microbe.

As Kendall probably recognized, the advent of a new technology such as the Rife microscope offered the possibility of technical closure to the controversy—at least the filtrationist side of it—by making filterable viruses observable phenomena rather than theoretical postulates. He was only wrong to the extent that he choose the Rife microscope rather than the electron microscope. (Indeed, because the Rife microscope allowed for observation of living phenomena, like other darkfield microscopes it would have favored the pleomorphic interpretation.) Later, when the advent of molecular biology made genomes more the principle marker of species categories than morphology, nutritional parameters, and staining patterns, it became easier to tolerate variation. The study of cell-wall deficient bacteria and pathology was able to return to life without being so threatening. In today's postmodern world of shape-shifting cyborgs, flexible accumulation (Harvey 1989), and what Emily Martin (1994) calls "flexible bodies"—where cultural and genetic diversity are often positively valued and standardization is associated with cultural hegemony—it is easier to imagine more flexible classification systems for microorganisms, systems that would make even Hadley's and Glover's bacterial life cycles seem standardized and unduly stabilized (see Domingue 1995, 1996).

The Gendering of Microbes, Researchers, and Therapies

So far my application of the culture concept has been restricted to identifying the orthodoxies and heterodoxies of the emerging professional cultures of cancer research and bacteriology, and their relationships to the values and interests of industrial culture. I have shown how consensus cultures emerged in cancer research and bacteriology, and how those cultures were consistent with professional and financial interests, as well as the general modernist cultures, in which they were embedded. The next step synthesizes an interests analysis of the morphology of research cultures with an analysis of gender that is rooted in feminist theory. This step implies more than merely adding a layer onto the interests theory or another dimension to the analysis of the research cultures. Instead, gender transforms an interests-and-culture analysis into a culture-and-power analysis. Interests are not forgotten; for the analysis of gender involves recognizing the material interest of men in maintaining for themselves positions of prestige and power. However, gender also allows for a rereading of interests theory to show how the instrumental aspects of protecting one's sacred theories or material position are complicated because they are suspended in a network of more general cultural values and categories. This argument will proceed in two steps: to demonstrate the gendered logic at work in the culture of microbiology, and to demonstrate the gendered logic in the culture of orthodox cancer therapies.

In the case of microbiology, the relatively marginalized position of research on bacterial variation, when viewed through the lens of studies of gender and social stratification in science, would suggest that the field of research on bacterial variation might have a relatively large number of women, or at the minimum to be perceived to have a large number of women.[40] This is indeed the case. In the first (1974) edition of the textbook *Cell-Wall Deficient Bacteria,* microbiologist Lida Mattman introduces the topic of the role of women in the concluding chapter, which she titled "L-Variants and the XX Chromosomes." In the second edition, published in 1993, Mattman retitled the chapter "The XX Chromosome." Because the chapter in the first edition more explicitly raises some interesting gender issues, I will focus on Mattman's discussion there.

The term "L-variants" or "L-phase variants," which appeared in the title of the chapter on gender in the first edition, refers to cell-wall deficient bacterial colonies that look like fried eggs and are capable of reverting back to classical bacteria. (The term "L Forms," which today is one key word for picking up research on cell-wall deficient bacteria in Medline or other research databases, was coined by Emmy Klieneberger-Nobel. She named the bacterial variants after the Lister Institute, where she worked.) In the chapter on L-Variants and the XX chromosome, Mattman comments on the large number of women who were doing research on cell-wall deficient bacteria. In an appendix to the book, she provides a portrait gallery of the many women who contributed to the field. She offers her own psychological explanation for the prominent role of women: "Working with cell-wall deficient forms requires great patience, a characteristic usually considered feminine. Or, facetiously, one might state that following cultural growth by the microscope is a sedentary occupation and thus appeals to women" (1974: 386).

Microbiologist Kenneth Bisset uses a similar explanation in his comments on the relationship between difficulty and controversy: "Work on these fragile organisms [L-forms and mycoplasma] is not easy, and it has been highly controversial . . . partly for the simple reason that it *is* so technically difficult." He adds that this is among the main reasons why "it is the courageous sex that has persevered at the task, almost unsupported" (1969: 581). Yet, technical difficulty alone is not enough to explain the low status of the field. A full understanding requires reference to the lingering effects of the various controversies over filtrationism and cyclogeny, as well as the purported lack of practical value of the research because many researchers do not think of CWD organisms as pathogens. Together with the technical difficulty, these features can explain the low status of the field. Because men tend to occupy the higher status fields, low status can explain the preponderance of women, or at least the perception that the field belongs to women. This explanation seems more appropriate than the psychological theory found among microbiologists such as Mattman and Bisset.

A related issue to the role of women in the subfield is the gendered cultural meaning of the research. Mattman provides a starting point with the following passage:

> Women who persist in L-variant research after marriage may find themselves subject to open slander. They may be accused of serving shish kebab to the family, made of protoplasts and spheroplasts alternately arranged on a skewer. Likewise, such female investigators, if they have small children, are suspected of starting bedtime stories with the phrase "once upon a time there was a poor little bacterium who had lost his cell wall." (1974: 387)

Mattman is clearly referring to some aspects of the sexism prevalent in science, a problem that was much worse—or at least more overt—for her generation than for younger ones. At the risk of reading too much into her comments, the symbolism should be underscored for the psychoanalytically unattuned reader. The image of a gendered "poor little bacterium" who had lost *his* cell wall, like a little boy who has lost his phallus, suggests a sexist comparison of CWD bacteria and women researchers as deficient. This comparison is even more inviting when one realizes that Mattman prefers to think of CWD as meaning cell-wall "divergent" rather than deficient. As she notes, "the term 'deficient' implies fragility and other labile traits which are not always present" (1974: 9).

Whether or not Mattman intended the comparison between women microbiologists and cell-wall divergent bacteria, let us make it explicit here: women, like CWD bacteria, were capable of surviving, and even prospering, under conditions that would be impossible for most well-connected male scientists. Working in the interstices of the funding system, or sometimes inside the laboratories of men, they patiently persevered with their work for decades. Research on bacterial variation continued throughout the twentieth century, but in a relatively hidden and unrecognized state, much like cell-wall deficient bacteria may persist in a latent state inside host cells. In today's world—where diversity, variety, and flexibility are valued attributes of organizations and people—the long-hidden research tradition on stealth pathogens may be making a comeback, much like latent bacterial pathogens that reappear in a host with a compromised immune system or nutritional status. In a similar way, the hidden research tradition on CWD bacteria—one which has an apparently marked prominence of women researchers—may also gain some recognition.

The general point is that in the culture of microbiology the controversies over variation have a richer meaning than merely the defense of a core theory of modified monomorphism that was set in place during the modernist era. The research tradition on bacterial variation is not only marginalized but gendered. Standardized and cell-wall divergent bacteria are technototems that mark off cultural spaces within microbiology, just as they have historically tended to mark a division of labor that symbolically linked women with bacterial variation. Cell-wall divergent bacteria are microbiological deviants, just as Hadley, Glover, and Livingston were microbiology and cancer-research deviants. Even today CWD bacteria are still considered aberrations against the norm of standard bacteria with cell walls. Lacking a cell wall, these bacteria are presumed to be defenseless, and they are frequently assumed to be harmless and fragile (or passive) because they are supposedly destroyed under osmotic pressure. The gendering of bacteria has accompanied the relative lack of recognition of their possible role in chronic disease.[41] As a result, opportunities have been lost for a better understanding of disease, perhaps even important diseases such as interstitial cystitis, arthritis, and cancer.

A similar argument regarding the permeation of gender values can be made for cancer research. An outstanding aspect of the research on bacterial variation and cancer in North America, particularly after World War II, was the prominence of women's network within the field. Virginia Livingston, Eleanor Alexander-Jackson, Irene Diller, and later Florence Seibert all took on a leadership role in this field from the 1950s through the 1980s. Seibert's autobiography *Pebbles on the Hill of a Scientist* points to the many women scientists she knew, and it seems that part of the legacy she wished to leave behind was a model for women in science. Likewise, Livingston never explicitly identifies herself in her writings as a feminist, but when I discussed this question in 1995 with Patricia Huntley, the director of the Livingston Foundation, she stated that Livingston was clearly a feminist in deed and thought if not in explicit label. Livingston's writings show that she was clearly aware of the prejudices that she experienced as a woman. As early as her medical-school years and her struggles as a resident among the venereally infected prostitutes, she developed an acute awareness of how she was working in a world controlled by men.

As Livingston's career developed, her networks and allies included men, but clearly women were crucial and prominent in her research efforts, particularly Alexander-Jackson and Diller. Several of her early publications were in the *Journal of the American Women's Medical Association,* and researchers sometimes asked her what she had published because they did not read the journals in which she published.[42] Livingston became an example of what Margaret Rossiter (1993) has called the Matilda effect: the tendency for women to suffer from under-recognition and cumulative disadvantage feedback loops in their career trajectories.

Although Livingston did have affiliations with Rutgers University and the University of San Diego, her career did not follow the standard pathways of a university researcher. She worked as a school doctor for some time, and she did not have powerful networks that could protect her when needed. When she went to work in a clinic in San Diego, she was acutely aware of her position as the last physician hired and as an "older woman" (1972: 81). She notes that although she received a man's salary she had to work much harder for the pay, often while the men were gathered in their offices smoking and talking about baseball. She writes, "When I would suggest that they might like to take some of my overflow, the reply was, 'You don't smoke and you don't know anything about baseball. You're not one of the boys' " (81). Later, she adds, "When I became well-known at the clinic, I was a threat to their prestige" (85). Without a doubt the fact that she had to work as a clinician at various points of her career took its toll on her ability to produce research. By most accounts of people who knew her, Livingston was a stubborn and arrogant individual, but her colleagues rapidly add that those traits were necessary for her success and survival in a doubly uphill battle as a woman medical researcher and a doctor who advocated alternative cancer therapies.

Some of the images in her autobiographical stories also give a sense of her experiences of researching as a woman. She writes of the sadness of extracting tissues from recently removed breasts, which were ideal for their work because the cancers were usually enclosed in the breast and not contaminated as in the case of bowel cancers. In contrast, one of her occasional moments of gendered humor appears in a description of an adventure with foreskin tissues. She found that the foreskins of circumcised babies were a good source of healthy human tissue for culture:

> So on an icy, cold, dark February morning, our tissue culturist, Marilyn Clark, went over to collect the morning's harvest of fresh foreskins. I glanced out of the window of the tissue culture room as Marilyn started across the icy courtyard carrying the neatly covered tray of foreskins. She was cold and she was hurrying. All of a sudden her feet flew out from under her and down she went, scooting along the ice. The tray flew out of her hands, scattering foreskins into the air. Scarcely before they had a chance to reach the ground, a flock of sparrows hunched on a neighboring telephone wire, swooped down and in a twinkling, all the little pink and white delectable foreskins were snapped up much like the pink worms that appear on a grassy lawn after a summer shower. Marilyn got up, brushed herself off, picked up the tray and gathered up the gauze. When I saw that she was not hurt, I was convulsed with laughter. The thought that those cold, hungry little sparrows were unexpectedly treated to a delicious meal cheered me up for the rest of that gloomy day. (1972: 39–40)

Another instance of her humor is her protest against the dream of Dr. John Lawrence of Lawrence Radiation Laboratories (later Lawrence Livermore Laboratories), which he announced during the 1966 seminar for science writers sponsored by the American Cancer Society. Lawrence called for "a cobalt machine in every town and village in the United States" (Livingston 1984: 94). In the context of the common linguistic memory of most Americans, the phrase "cobalt machine in every town" is likely to bring to mind the old campaign phrase "a chicken in every pot." In later chapters including the one titled "Chicken: Cancer in Every Pot," Livingston discusses the Rous virus and the dangers of cancer crossing species to humans who eat undercooked poultry. She clearly preferred her immunotherapies and dietary therapies to the cobalt machine, which Rhoads had tried to force on her after he had taken away her grant. As she comments, when Lawrence made his statement about a cobalt machine in every town, "I became excited and began to wave a petri dish (used for making cultures) over my head and said that it could be mightier than all the high-powered radiation machines in the world" (94).

The contrast of a machine that zaps cancer versus a dish that cultures microorganisms is implicitly gendered, and here we pass from the individual experience of Livingston as a woman researcher to the gendered contours of the culture of cancer research. One might argue the case psychoanalytically based on homologies between radiation machines and guns versus those between petri dishes and cooking (or phallic versus vaginal symbols), but there is a more direct way of understanding the gendered opposition between most orthodox cancer therapies and many alternative ones. Indeed, once one begins to see gender in the cultural construction of cancer therapies, the whole divide between the nutritional, nontoxic, and nonspecific immunotherapies on the one side versus the conventional cancer therapies of "slash, burn, and poison" shifts from simply a case of interests, in which the everyday, common-person of alternative medicine is paired off against the big money of the medical and pharmaceutical establishment. Although many researchers have seen this struggle as a class or economic division, the gender dimension of the imagery of "warfare" in the war in cancer has remained underexplored.[43]

Cancer researchers often signal their aggressive approach to cancer treatment through metaphors. Moss, following Irwin Bross of Roswell Park, notes that military metaphors are deeply ingrained:

> We have now become accustomed to talk of "weapons," "strategies," and a whole "armamentarium" of cell-killing drugs. Combinations of drugs bear aggressive-sounding acronyms like BOLD, CHOP, COP, COP-BLAM, ICE, MOP and ProMACE. (1995: 22)

Furthermore, the aggressive and heroic values in the culture of cancer treatment are particularly strong in the United States. American medicine has long been recognized as having a relatively aggressive cultural style, particularly in cancer therapy. In *Medicine and Culture* journalist Lynn Payer writes:

> Even as Europeans were developing the simple mastectomy and the lumpectomy as less mutilating ways to treat breast cancer, American doctors were advocating the super-radical mastectomy and prophylactic removal of both breasts to prevent breast cancer.

> American medicine is aggressive. . . . American doctors perform more diagnostic tests than doctors in France, West Germany, or England. They often eschew drug treatment in favor of more aggressive surgery, but if they do use drugs they are likely to use higher doses and more aggressive drugs. (1988: 124–25)

One might counterargue that women with breast cancer have come a long way since Payer wrote this passage in 1988, but that progress is largely due to a great deal of hard work and organizing by many people, including many women patients. Certainly medical opinion in the United States has shifted on the question of radical mastectomies. Yet even as use of the radical mastectomy has become more limited, aggressive use of chemotherapy has tended to fill the void (Moss 1995: 89).

The aggressive and heroic treatment of cancer in the United States has elective affinities, to use the term Max Weber (1958) brought into the social sciences, with other major value complexes in the general culture. The metaphors and value system that undergird an aggressive approach to cancer are drawn from the construction of evil in the American cultural tradition. By deploying the warfare metaphor, Rhodes and Nixon transformed, respectively, the World War II rhetoric about the fascist Other and the Cold War rhetoric about the communist Other into a powerful image of cancer as a dangerous and evil Other. In turn the warfare image is supported by the Calvinist religious heritage that views alterity not as a difference to be tolerated but as an evil to be eradicated, defeated, or driven out of existence.[44] The final linkage of evil Others that must be destroyed shifts the chain yet one more time: from Satan to Nazis (or Communists) to cancer to the cancer quacks. When the quackbusters and cancer-establishment researchers (in contrast with the often patient-sensitized clinicians) pronounce on alternative medicine, their self-righteousness is both an expression of a sincere, quasi-Calvinist belief in the sacredness of science and of a more general cultural logic that tends to polarize the world into good guys and bad guys (DaMatta 1991). Medicine wraps itself in the sacredness of science much as the political culture legitimates itself through reference to the masculine, Calvinist God.

Markings of male and female characterize not only the boundary between scientific medicine and its Others that are unproven, unconventional, alternative, adjuvant, or complementary, but also the various boundaries within the domain of orthodox medicine. For example, one of the top subfields, if not the peak, of the medical profession is surgery, a field that is historically as masculine as nursing, pediatrics, nutritional counseling, social work, or other lower-status fields are feminine. Surgery is also the heart of conventional cancer therapy. The surgeon's relationship to cancer is a masculine one: he defeats the foe by cutting it out and removing it, not by nutritionally or immunologically rebalancing the body. In the United States the surgeons at Memorial Hospital and at Johns Hopkins set the tone for the aggressive, masculine style in the American treatment of cancer. For example, in the 1890s William Halsted of Johns Hopkins inaugurated the American tradition of the radical mastectomy, and some of the subsequent surgical procedures at the Memorial Hospital were even more heroic. Their procedures included the infamous hemicorporectomy—removal of the lower half of the body (Moss 1989: 49).

Surgery was also gendered in a second way: its tendency to be directed disproportionately at women and especially at women's diseases. Another founding father of cancer surgery, J. Marion Sims, who triples in the history of American medicine as the father of modern gynecology and also a founder of the first private cancer hospital (the predecessor of New York's Memorial Hospital), remains controversial even today for the excesses of his work.[45] The overuse of surgery for women's diseases of all sorts is a well-established finding in the history of medicine, especially for the earlier phases of psychiatry that predated psychodynamics and the talking cure. In comparison with psychiatry, the aggressive use of surgery has only been challenged much more recently for cancer and related diseases such as fibroid tumors.

Of course, surgery—even radical surgery—survived as a practice not only because of its cultural meaning but also because of its efficacy. There is no denying that surgery provides a valuable option in some cases, particularly early stages of solid tumors. A key study from the alternative community demonstrates that for melanoma patients surgery combined with the Gerson dietary therapy is more efficacious than the dietary therapy alone (Hildenbrand et al. 1996). However, interwoven in the referential and pragmatic question of efficacy is a series of often unquestioned cultural assumptions about the nature of cancer, assumptions that color attempts at scientific evaluation of alternative therapies. At the root of the culture of conventional cancer theory and therapy, from at least the late nineteenth century to the present, is the idea of eradicating the enemy, of getting it all, with all of its resonances in warfare, masculine heroism, and a quasireligious Evil Other. Radiation and chemotherapy, first introduced as adjuvant therapies, continue this cultural logic of the cytotoxic therapy that aims to get it all. Even the more recent immunotherapies are constituted as specific in the sense that they target cancer cells like hi-tech missiles, and genetic therapy is presented in similar terms. For example, a headline from 1996 reads, "Genetic Time Bomb is Designed to Deliver Killer Force to Cancer Cells."[46] Rhoads's cancer war continues in the age of molecular biology.

The alternative approach, which can and should be constituted as a modality that complements a more perspicacious use of surgery, focuses on changing the metabolic or nutritional environment of the cells and on strengthening the immune system. From this viewpoint Rhoads's metaphor of treating the root cause of cancer, the "soil" in which it grows, appears to be similar the metaphors of the alternative metabolic and nutritional therapies, which propose controlling cancer by rebalancing the nutrients and metabolic processes of the cell and its environment. However, Rhoads's soil metaphor got swallowed up in his metaphor of warfare and his concern with developing chemotherapies that could search out and destroy cancer cells. To continue the soil metaphor, Rhoads was advocating pesticide control, not organic farming.

Livingston and other advocates of nontoxic cancer therapies focus not on destroying tumors but on building up the body through diet and nontoxic immunotherapies such as bacterial vaccines. For her the nutritional status of the body was closely linked to the immune system's ability to control microbial infections. Her book *The Conquest of Cancer* ends with an eighty-page collection of recipes for the cancer-free diet. There is now a massive amount of research and a widespread consensus that diet and lifestyle are crucial variables in the risk factor equations for cancer prevention. Yet, as treatments they are gendered and marginalized from the perspective of a therapeutic ideology that posits heroic, masculine methods of destroying cancer, not to mention a general culture that has long associated cooking with the female. From this perspective, to the extent that nutrition is recognized as having a role in cancer therapy, it is as an adjuvant treatment to cytotoxic therapy, thus continuing a gendered cultural logic of supplementarity (Culler 1982).

In short, bacterial approaches to cancer were marginalized for more complex reasons than financial and professional interests or even the modernist preferences for standardization. Bacterial approaches became caught up in a gendered cultural logic in which both the object of analysis, pleomorphic microbes, and the nontoxic therapeutic approaches of nutrition and vaccines were gendered as female. In a heroic culture of microbe hunters who achieved places in the scientific hall of fame because they could capture and display stable microorganisms, and of surgeons and oncologists who made promises of doing battle with the cancer foe and getting it all, the alternative approach resonated with sexist views of undesirable female traits. Bacterial approaches to cancer, like other alternative therapies, seemed destined for the dustbin of scientific and medical history . . . except for one small variable: patients.

Cancer in a Multicultural World

The final aspect of a cultural and political interpretation of cancer research to be discussed here involves the complex issue of globalization and international flows of patients in search of cancer cures. Including this level of analysis is crucial for two reasons: it reveals the limitations of current policies, such as the quackbusting strategy, and it further develops an analytical framework for an explanation of medicine in terms of culture and power. From this perspective, cultures are not merely contrasting and sometimes conflicting bodies of assumptions, values, and ways of being in the world, but they are resources to which people may turn in their efforts to maneuver around the mechanisms of suppression.

At the time when the microbial theory of cancer was rejected, there was no phenomenon of organized patients' advocacy groups and referral organizations that shared information and contested the knowledge received from the leaders of the medical profession. Patients lived in much more localized worlds; their information was more limited, and the ability to travel to other countries for different therapies was much more narrowly limited to the rich. The globalization of the world that has occurred during the twentieth century has crucial implications for medical consensus and the role of the public in shaping it. Globalization is accompanied not only by increased diversity of opinions, but by increased access to information and alternative therapies themselves.

The medical profession's repressive strategy for managing the growth of alternative medicine, which is still in force today, is a product of the modernist period. I like to think of the strategy as a sociological equivalent of tyndallization. In science and technology studies, the English scientist John Tyndall is known as an exemplar of what Thomas Gieryn (1983a, 1983b) has called boundary-work, that is, attempts by scientists as public figures to distinguish science from other cultural domains in the effort to gain prestige, autonomy, and funding for their activities. As professor and superintendent at the Royal Institution in London and president of the

British Association for the Advancement of Science, Tyndall used his credentials to improve the position of science with respect to the two competing cultural domains of religion and engineering. In the process, he debunked popular superstitions in ways that are reminiscent of today's skeptics and quackbusters.

In biology Tyndall is remembered for another reason. When attempting to replicate Pasteur's work, he found that bacteria continued to exist in heat-resistant forms that could survive pasteurization. He found that repeated heating on various days was necessary to destroy the endospores. Consequently, tyndallization became a more powerful form of pasteurization. A metaphor of tyndallization—of repeated sterilization—seems appropriate as a description of the cancer establishment's repressive strategy toward alternative cancer therapies.

However, the repressive strategy has becoming increasingly ineffective. Because cancer rates continue to grow and because long-term survival rates have shown little if any improvement, contemporary culture is a fertile terrain for the growth of alternative cancer therapies. It would probably be impossible for the government to stop all alternative cancer treatments; like microbes, they seem to be omnipresent in the environment if not latent like the intracellular pathogens. Instead of providing complete repression, the legal apparatus is used strategically and repeatedly, like tyndallization. In other words, repression is reserved for those cases that gain enough public attention to reach a critical level of visibility, such as Rife or Livingston. Once visible, they become examples for the rest of the movement (thus reinforcing the value of maintaining a low profile) and to the public, which is treated to sanctimonious lectures on quackery that are reminiscent of the old Puritan tales of witchcraft. The many stories of the persecution of alternative doctors and scientists—and I have only scratched the surface in this examination of one research tradition—are widely circulated in the books, articles, and conferences on alternative cancer therapies. Those stories have a warning effect on others who might wish to become active in the area.

However, because tyndallization occurs in an open culture that is internationalized and media-saturated, the old strategy is increasingly ineffective and even counterproductive. Alternative therapies continue to proliferate. Some practitioners may be cowed by the repressive strategy, but the end result in the contemporary world is likely to be the opposite. Stories of repression are great grist for the media mill, and the media tend to present controversies in a neutralist vein. As we known from the science-studies literature, neutral accounts tend to be captured by the out-group (Scott, Richards, and Martin 1990). In short, the tyndallization strategy may end up advertising alternative medicine as much as repressing it.

The repressive strategy also suffers from a quantitative problem. New ideas and new therapies spread even more rapidly than in the past. Cancer patients have benefited from the work of the much-better-organized AIDS movement and also from the burgeoning New Age movement. There is a growing supplements industry that finances scientific research on nutritional supplements and popular magazines to relay this research to the public. The literature on alternative cancer therapies is readily available in New Age bookstores, health-food stores, and even in some of the health sections of chain stores. Numerous institutions provide information, newsletters, and conferences. Although labelling laws prevent the linkage of health claims to packaging of nutritional supplements, consumers are only slowed by this attempt to prevent the free flow of information, and they find the information they need in books, magazines, and newsletters. By the middle of the 1990s, it was also becoming increasingly easy to find this information on-line. In short, the repressive strategies that worked during the peak years of the modernist period—the middle decades of this century—operate today on a movement that has changed. Like the new antibiotic-resistant microbes, advocates of alternative cancer therapies today are much more resistant and resilient.

Globalization has played an important role in the resilience of alternative medicine. One way in which alternative medical practices can escape the strong arm of state-sanctioned repression is by moving across cultural borders. While the leaders of the cancer establishment and quackbusters still act as if they were seventeenth-century Puritans, who attempted and failed to establish a religious state with no toleration for sectarian diversity, Mexico and the Caribbean serve as the new Rhode Island and Pennsylvania. A dynamic model of how alternative medicine works today in the United States (and perhaps in some other highly regulated countries) must take into account the relationship between globalization and the growth of medical pluralism. Cultural borders provide a countervailing force to the repressive strategy because they can serve as resources for legal end-runs. In the context of the law the term "loophole" might be used as an approximate translation of a well-understood concept in Brazilian culture known as the *jeitinho,* that is, the ability to bend rules and figure out a way around official restrictions (Barbosa 1995, DaMatta 1991). Often, however, the *jeitinho* implies going outside or around the law, in the sense of the American football metaphor of the end-run around the wall of players constituted by the defense.

One of the ways in which alternative medicine survives is by crossing cultural borders and finding a new home in other countries. Medical regulation now operates in a globalized world in which national frontiers are increasingly permeable. Consider a few examples from the research on bacteria and cancer that I have examined in this book. In recent years Coley's toxins have begun a comeback in the international context. Although clinical trials and uses of Coley's toxins have been all but abandoned in this country, they have found a second life in the People's Republic of China. Dr. Guo Zheren established the Coley Hospital in 1990, and he and another Chinese researcher have published outcomes research on their successful use of the Coley's toxins (Zheren and Nauts 1991; Tang et al. 1991).

Another example of a researcher who accepted the microbiological theory of cancer in some cases was William Frederick Koch. He developed an oxidation product known as gloxilide, which according to his former "disciple," Dr. Jayme Treiger, was "based on the correct activation of the carbonyl groups existing in the cells, through the creation of an intracellular chain reaction to neutralize pathogens."[47] Koch was aware of the work of the bacteria-and-cancer researchers, and he believed that the beneficial free radicals of his product could destabilize cancer pathogens. In the 1950s Koch left the United States during a period of prosecution by the FDA, and he relocated in Brazil. His therapy continues to be used in Brazil by Treiger.

By far Mexico has been the greatest haven for alternative cancer therapies. The 714-X drug of Naessens, who left France to live in Quebec, is available legally at Mexican clinics in Vera Cruz and Tijuana. Bacterial vaccines are being used in the Centro Hospitalario Internacional Pacifico, SA, itself a product of the suppression of Max Gerson in the United States. Likewise, Dr. Geronimo Rubio of the American Metabolic Institute in Tijuana uses a Rife machine as part of an overall program of immunological treatment for some of their patients. Among the immunotherapies offered by the hospital are autogenous vaccines for cancer viruses that are similar to the Livingston vaccines.[48]

The Tijuana cancer clinics and hospitals are perhaps the most well-known example of this ability for therapies to survive by crossing cultural borders. The phenomenon has reached such a point of institutionalization that the Los Angeles–based Cancer Control Society hosts annual meetings where representatives of the clinics speak along with many other leaders of the alternative cancer-therapy community. Patients who are interested in selecting a Mexican clinic are able to take bus tours of the Tijuana cancer clinics.

For anthropologists the transnational flows of patients, clinicians, research, and therapies are old hat: contemporary anthropology no longer sees cultural borders as impermeable walls across which comparisons are made. The multiple international flows of people and practices in the global village have been catalogued in general in the well-known work of Arjun Appadurai (1990) and for science in the work of Sharon Traweek (1992). However, the transnational flows of the alternative cancer movement may also have something to teach social scientists about culture in the globalized postmodern world. In some cases, cultural borders are less boundaries to be transgressed by transnational flows than resources that actors can use strategically to alter power equations in negotiations over scientific legitimacy and medical legality.

As a resource, globalization works for both alternative medical practitioners and their would-be regulators. These days one does not simply disappear from the panoptical gaze of the American government's regulatory apparati by relocating to the Caribbean or Mexico. For example, in the 1970s the New York–based doctor Lawrence Burton, known for his immunoaugmentative therapy, set up practice in the Bahamas after the FDA blocked Investigational New Drug status. He practiced unhindered for a while, but in 1985 a patient brought vials of immunoaugmentative therapy serum from the Bahamian clinic to a blood bank in Washington state, where tests of the samples indicated that they were positive for the AIDS virus. The claims were probably based on false positives from the nonsensitive ELISA test, and subsequent testing by the more sensitive Western blot test and of over fifty Burton patients indicated no evidence for seropositivity. Furthermore, in the mid-1980s the U.S. blood supply was widely contaminated with HIV antibodies, and the Burton clinic was probably being singled out for political reasons. However, the scare campaign had its effect: the Bahamian Ministry of Health closed the clinic temporarily.[49]

The Burton case demonstrates that even when clinics move offshore, they may still have to negotiate with the regulatory arm of American medicine and science. However, countries such as Mexico, the Bahamas, and Brazil tend to regard as imperialism any pressure from the U.S. government or U.S.-influenced international organizations, and they may react negatively to outside interference. The end result at this point for the Burton clinic is interesting: notwithstanding the reach of the American state into the Bahamas, the clinic continued to operate after a brief closing.

Moreover, alternative medicine is taking its place in the construction of national identities. In Brazil when I did anthropological fieldwork among the Spiritists, I constantly encountered critical references to the "orthodox medicine" or "orthodox science" of the rich countries (Hess 1991). From this perspective, politicized leaders of poorer countries see their embrace of alternative American medicine less as accepting cross-cultural medical dumping than as showing critical, anti-imperialist independence. Whereas both advocates and skeptics of alternative medicine in the rich countries may see the offshore clinics as legal end-runs, their advocates across the border often view the same maneuvers in different terms. In other words, support for unwanted alternative medicine in the poorer countries can mean showing the superior open-mindedness and freedom of their cultures, in stark contrast to Yankee rhetoric about its international role as the land of the free.

In Mexico, for example, the leaders of the alternative clinics see the relatively deregulated Mexican medical system not as a sign of inferiority but as a source of medical freedom. In Brazil the doctors with whom I spoke in 1995—both orthodox and un—were well aware that the chaotic regulatory apparati of their government, medical syndicates, and insurance providers meant that they had much more medical freedom than their counterparts in the United States. Of course, lower levels of regulation open the door to charlatanism and fraud, which are more common in Brazil, but the Brazilian doctors I knew were not willing to trade their medical freedom for the highly regulated environment of the United States. The point is that as alternative medicine becomes globalized, the level of regulation in other countries, particularly in Latin America, is unlikely to match that of North America and Europe. The appropriate policy response to the de facto deregulation implied by globalization is not to try to negotiate some kind of international standard (which will be left honored on paper only in countries like Brazil), nor to heighten the antiquackbusting rhetoric (which will only advertise alternative medicine), but to support publicly funded research and education on the relative efficacy of alternative therapies. In other words, the solution is not tyndallization but evaluation: test the alternatives, test them fairly, and let the public know the results of the tests.

To conclude by way of the metaphor of cell-wall divergent bacteria, from the perspective of orthodox medicine and the regulatory science of large, randomized, controlled, prospective, multisite, double-blind, cross-over clinical trials in the rich countries, alternative cancer therapies are constructed as deficient. In most cases alternative cancer therapies lack acceptable empirical support in the form of animal experimentation and clinical trials, and usually they also lack officially sanctioned legal status. To survive in this milieu, the alternative medicines, like filterable bacteria, shed their local organizational encumbrances and cross the semipermeable boundaries of national frontiers. As occurred with slave religions when imported into the New World, in the new culture the alternative medicines may regain their institutional settings and clinical functions (Bastide 1978).

The complexities of the co-construction of national, legal, and scientific borders provides a lesson for theories of science, technology, medicine, and society. In the case of alternative medicine (as in other kinds of heterodoxy), it is entirely inadequate to remain inside frameworks that see scientists as entrepreneurs who achieve success by building networks, accumulating symbolic capital, or serving occupational or class interests. Instead, culture is all around us: in the formation of research cultures with their negative heuristics and commonsensical doxas, in modernist and postmodern approaches to classification and standardization, in the gendering of microbes and cancer therapies, in cross-cultural variations of legal and institutional apparati, in the *jeitinhos* found in response to the regulatory disciplining, and in the pervasiveness of shared and contested understandings of ethnic and national identity. Much as Pasteur on his deathbed is said to have said, "The terrain is everything," so I would end this discussion of science, medicine, and society by advocating a thicker appreciation of the role of cultural borders that, like cell walls, operate as resources that can be built up, torn down, or transgressed on a strategic basis.

The implications of cultural borders, however, are more than theoretical. For tens of thousands of cancer patients, cultural borders may mean the difference between life and death. If their right to medical freedom of choice is denied in the United States, many have opted to go outside their home country where they can encounter the therapies that, in their evaluation, are worth trying. The situation is similar to the flight of women to other countries during the period prior to *Roe v. Wade* in the United States. In both cases medical freedom is a privilege of those who have the financial and intellectual resources to pursue it. In effect medical freedom becomes not an explicit, *de jure* right of all citizens but an implicit, *de facto* right of the wealthy. The situation is in sharp conflict with the general value of equality. The solution is to evaluate the alternative theories and therapies, and to change policy so that the public's right to medical freedom is guaranteed. At this point description ends and prescription begins.

4

But Is It Good Science?

So what about the "science"? Is it credible? Let us begin the question of evaluation with a clear definition of what is being evaluated. Although I have classified this research tradition as falling under the general theory that bacteria are causative agents in cancer, it should be clear that some of the work would be better termed microbial. Rather than thinking in terms of pleomorphic bacteria, some of the researchers advocated the existence of a new type of microorganism that has features similar to fungi, bacteria, viruses, and (in the case of Enderlein) even something approximating prions. However, in the English-language literature of the second half of the twentieth century, the organism is conceptualized more as a bacterium that has a filterable form as well as funguslike forms similar to the mycobacteria. The point is that what I will call the bacterial theory of cancer can be distinguished from other infectious theories of cancer. The other theories include not only the work of tumor virologists but also that of relatively unknown researchers who proposed that parasites, yeast, or amoeba—rather than a pleomorphic bacterium—play an etiological role in cancer.[1]

In addition to competing theories that assume some other type of microbial infection, the more conventional point of comparison is the secondary-infection theory. Under this null theory, bacteria associated with tumors represent secondary infections that play no role in tumor genesis and little if any role in tumor growth. Note that either the bacterial or the secondary-infection theory could be encompassed by the broader molecular theory of cancer. This standard, mainstream theory holds that cancer emerges from genetic damage such as inherited defects and unrepaired lesions in the DNA of the host cell. Because the molecular theory represents consensus knowledge, it must be viewed as the encompassing theory. The comparison takes place at the level of two subtheories: are bacterial infections unimportant to tumor genesis and development, or do they represent an overlooked factor in the cancer process?

Guidelines for Evaluating a Theory

Because the problem involves comparing theories (and their affiliated research programs), it follows from work in the philosophy of science that there is no simple crucial experiment or falsifying instance that can refute one of the theories beyond doubt. One can always come up with subsidiary ad hoc theories to explain the contrary results and anomalies. Likewise, there is no algorithm or computer program into which someone can enter the data and come out with an answer that determines whether the bacterial-etiology theory or secondary-infection theory is better. Evaluating a theory is a much more complicated process. It is more like a court trial or a medical diagnosis than a mathematical proof; distinctions between true and false, or good and bad, can be made, but they lack the exactitude of a mathematical formula. There are only guideposts that can suggest whether the theory is worth pursuing or not.

Elsewhere I review a number of positions in the philosophy of science and focus on the guideposts they suggest for theory choice (Hess 1997). Here I apply those ideas by dividing theory choice criteria into four major groups that I term positivist, conventionalist, pragmatic, and feminist. This set of criteria makes it possible to provide a fair and balanced evaluation of the theories by examining issues such as evidence, theoretical consistency, therapeutic potential, and possible social biases.

The first group of criteria focuses on the question of accuracy or evidence, and without a doubt it is the most important. The power and attraction of the game of science resides in its ability to allow evidence to play an important role in resolving many disputes. As Rudolf Carnap (1995) argues, a fundamental starting point is that the better theory explains more facts and makes better predictions. The first part, empirical subsumption, is not always easy to determine because the hard cores of theoretical systems may not be exactly comparable, and thus exact empirical subsumption may not be possible. In other words, in some cases the subsuming theory does not cover all the empirical research of the subsumed theory, a situation known as "Kuhn loss" (Fuller 1988: 223). Still, some form of evidence remains crucial to making a choice among theories or research programs. Recognizing the Kuhn-loss problem but also recognizing its limitations, let us say that the new theory should be at least empirically equivalent over a large range of crucial facts.

Most if not all the philosophers are also in agreement that the better theory should produce some new successful predictions. Thomas Kuhn's criterion (1977: 322) that a good theory be fruitful could be interpreted as another expression of the ability for good theories to produce new, even successful and surprising, predictions. Likewise, the realist criterion that over time the theoretical terms of the better theory will become observational terms (such as viruses in the history of microbiology) might also be included here as a subset of the accuracy group. In summary, there is a group of criteria that I like to think of as the positivist criteria in that they emphasize accuracy, either in terms of existing evidence or in terms of new predictions. However, I use the term "positivist" very loosely because I am including arguments about accuracy that have been made by nonpositivists.

I interpret the accuracy criterion to mean that the bacterial theory will have to be supported by observations that its competitor cannot explain. In turn, an advocate of the secondary-infection theory will have to present a large body of facts pointing to artifact and contamination that the bacterial theory cannot explain. The artifact counterarguments need to be considered in detail, because the accuracy argument could easily fall apart if counterarguments such as contamination can be convincingly defended.

In my opinion, the second most important group of criteria is what I think of as the conventionalist group. Following Pierre Duhem (1982), these criteria emphasize consistency issues. I also include simplicity and parsimony as part of the consistency group because these issues must be judged against a background of other theories and therefore against a general concern with consistency. However, Helen Longino (1994) has introduced another criterion, that a new theory be novel, which appears to be in conflict with the consistency/simplicity group. Her novelty criterion is a valuable reminder that the consistency group has conservative implications. For example, if scientists were to rely too heavily on consistency (at the expense of other criteria), they would find it impossible to accept revolutionary theories because these radical new theories were not consistent with consensus knowledge. A solution to the problem pointed out by Longino is to accept novelty when other criteria are met: the existing theory or research tradition has accumulated many anomalies or has not led to predictive or technological successes desired, the new theory is simple and parsimonious, and/or the novel theory restores theoretical harmony among apparently contradictory theories. In short, the consistency criterion is a very important one, but it must be used with some flexibility so that one is not trapped by its conservative implications and does not rule out revolutionary theories. However, the flip side of revolutionary theories is that they may turn out to be wrong, and this is why the other criteria—especially the accuracy group—are important.

Although many scientists and philosophers limit their evaluation criteria to some combination of my first two groups, I would suggest that two other criteria are very important. These criteria become particularly important when evaluating the more general issue of research programs, especially in the applied sciences, where questions of practical potential and social bias are more likely to be important. My third group of criteria emphasizes issues derived from the pragmatist tradition in the philosophy of science. To some extent the requirement of successful, surprising predictions falls into this group, especially when those predictions have new technological or therapeutic implications. Longino (1994) develops the pragmatic criteria when she suggests that a good new theory or research program be applicable to current human needs and that it favor "diffusion of power," that is, that the new theory be less limiting than a rival theory in terms of access and participation. Certainly, advocates of the bacterial approach believe that it has led to the discovery of new therapeutic applications that are relatively accessible and low cost. In practice, then, this group of criteria leads to an evaluation of the efficacy and practicality of therapies such as bacterial vaccines.

Finally, Longino (1994) has proposed theory-choice criteria that I group as feminist (or multicultural) criteria. One criterion, "ontological heterogeneity," refers to 1) a concern with diversity in the object of study, as in Barbara McClintock's attention to the diversity of the kernels of a corn cob (Keller 1985) or women primatologists' attention to diversity within and among primate groups (Haraway 1989); and 2) the rejection of theories of inferiority, that is, theories that see difference as substandard, deviation, or a failure. The second criterion, complexity of relationship, refers to the rejection of "single-factor causal models for models that incorporate dynamic interaction" (Longino 1994: 479). Bacterial pleomorphism would probably score highly on the ontological heterogeneity criterion, because it points to diversity in organisms where diversity has previously been overlooked. On the criterion of complexity of relationship, the bacterial-etiology theory would seem to reduce complexity by reducing cancer to an infectious disease. However, to the extent that the role of bacterial infection and disease is latent rather than contagious—and therefore stemming from overall systemic, nutritional, and environmental stressors rather than from exposure to carriers—a case could be made for complexity of relationship.

Although I find these criteria vague, I have included them here because there is an increasing diversification in both science and science studies (in terms of gender, race, and nationality), and consequently these criteria are likely to become increasingly legitimate to a large number of researchers over time. They are probably also likely to become more refined. I am not especially happy with these criteria as Longino has formulated them because they seem to be vague enough that one could probably find a way in which almost any theory satisfies them. As a result, I will propose an alternative criterion that maintains the spirit of their intent but does not suffer as much from the vagueness problem. This criterion is simply lower social prejudice than the rival theory (in terms of sexism, racism, etc.). This criterion may not be applicable to all theory choices in science, but there are some possible grounds on which it may be applicable to the specific problem discussed here.

In this chapter I will consider in some detail these four groups of theory-choice criteria as a way of evaluating the bacterial-etiology theory. As counterevidence emerges, I will point out ways in which the theory must be modified in order to pass the evaluation criteria. I suggest that, if properly modified, the theory may be able to build a reasonable defense on some of the criteria or groups of criteria that provide a ground for good theory choice.

Particularistic Criteria and Theory Evaluation

So far the four groups of theory-choice criteria—accuracy, consistency, pragmatic value, and low social prejudice—are universalistic criteria. In other words, they do not take into account the question of who is proposing the theory. Should any particularistic criteria be included in the evaluation? Some should be included as negative criteria, and in fact they routinely *are* included in peer-review processes such as grant review disclosure forms. In other words, if the evaluator can be shown to have a conflict of interest via favoritism or personal gain, one may tend to discount the evaluation. Because I have no personal connection with the advocates of the bacterial theory, no financial stake in any cancer theory, and no strong commitment to microbiological theories of cancer, I am relatively free from conflict of interest. In fact, I came upon this theory largely by chance (a book in a bookstore), and I found it so preposterous that I was curious enough to check some of the sources. Eventually this led me to rediscover several research traditions on the possible infectious component of chronic degenerative diseases.

Regarding the social-prejudice criterion, standpoint epistemologies would suggest paying particular attention to theories that are advocated by historically excluded groups (Harding 1992). Although application of this particularistic criterion would not affect theory choice directly, it would promise something like equal access to evaluation for theories coming from women and historically excluded groups. Relativism can easily be avoided because none of the other universalistic criteria are forfeited; one merely makes sure that these theories get a fair hearing. Of course, the problem is that this argument can be used to justify giving a fair hearing to every crackpot theory from every historically excluded social group. To avoid the problem, I use this criterion along with all the others to arrive at a complex judgement. However, I would argue that this criterion is relevant to the case of cancer research that I am considering here. The women's network of Livingston, Alexander-Jackson, Diller, and Seibert managed to achieve top-quality research during an era when it was particularly difficult for women to engage in science. At the minimum, this women's network in cancer research deserves a second glance.

Probably the most complicated of the personalistic criteria is reputation. The empirical literature suggests that reputation may be widely used as a criterion for theory choice or evaluation of lower-level empirical claims (Cole 1992). However, this is a descriptive claim; the prescriptive question is, *should* reputation be elevated to one of the prescriptive criteria for theory choice? On the one hand, researchers who have prestigious degrees, institutional positions, and journal publications can be expected to produce higher-quality research on the average. Reputation is therefore probably an economical index for making a preliminary cut between good and bad research. On the other hand, those who do not hold these markers of reputation tend to be the very outsiders who bring in badly needed new ideas, or they may be victims of the kinds of suppression described for the alternative cancer researchers. Keeping these two conflicting possibilities in mind, reputation has entered into my evaluation in the following way: when I selected which advocates of the bacterial-etiology theory to include in the evaluation, I decided to restrict it mostly to the Anglophone North American group running from Glover, Scott, and Clark to Livingston, Alexander-Jackson, Diller, Seibert, and colleagues. Much of the work of this American network is published in scientific and medical journals, for which there was peer review in some cases. As a result this research is the most "sanitized" of these heterodox research networks, to invoke Roy Wallis's (1985) term, and it is the most likely to provide a best case for the bacterial theory. If my evaluation finds the research lacking in this network, it is likely that the other research will suffer from the same problems.[2]

Problems with Béchamp

I will exclude Béchamp because of gross failure to pass accuracy and consistency tests. Nevertheless, because some writers in the alternative medical literature now present Béchamp as a lost hero in the history of medicine, it is worth pausing for a moment to address the question of whether or not he was right. In my opinion, most of what he said is plain wrong. Disease is often caused by contagious infection, and there are many bacterial, viral, and fungal pathogens in the environment. The mechanism of blood coagulation and the types of proteins involved are now textbook knowledge, and microbes are not part of the story. Béchamp was also an extreme pleomorphist; that is, he believed that the numerous species of bacteria could emerge from microzyma "by evolution." Even when one takes into account how the meaning of the word "evolution" has changed, we now have molecular technologies to determine the genetic differences that characterize different bacterial and viral species. It is possible that future science will uncover a world in which prions combine freely to form viruses, and viruses combine genomes to make bacteria, but so far the possibilities of genetic recombination seem more restricted.

Although much in Béchamp's theory is, in my evaluation, incorrect, he did propose a few ideas that can be seen as precursors of some ideas of the middle and late twentieth century. For example, it is now consensus knowledge that host cells are able to harbor microorganisms in latent states for long periods of time, and that these microorganisms are capable of causing disease when the host is weakened. A noncontroversial example is the herpes virus, but, as will be discussed shortly, there is growing research that suggests that some types of bacteria may exhibit similar patterns of latency and pathogenicity. Second, the idea that the blood of healthy individuals is not sterile is no longer dismissed out-of-hand. Although I will not evaluate the blood-borne microbial theories of Enderlein, von Brehmer, Villequez, and Naessens, subsequent research on bacterial blood pathogens confirms the general finding of microbial blood colonization in apparently healthy persons. For example, in the 1970s researchers at the University of Camerino Medical School published findings on staphylococci in the red-blood-cells of healthy individuals, and this work on red blood cell parasitism has been confirmed and extended by other laboratories, including that of Gerald Domingue, a microbiologist, immunologist, and urologist at Tulane University Medical School.[3] Third, the idea that at least some species of bacteria can undergo variation, including cell-wall deficient and filterable forms, is now an accepted part of microbiology. However, as far as is known today these changes are much more limited than those envisioned by Béchamp or Enderlein. Finally, the idea that there may be life forms smaller than viruses is now receiving attention in research on prions, some of which are now recognized as pathogens.

Although it may be time to grant to Béchamp and other nineteenth-century students of bacterial variation a larger role in the history of microbiology, particularly regarding some of the specific discoveries that antedated Pasteur's work, it is simplistic to view him as the fearless underdog who spoke the truth as science became pasteurized. Béchamp had some very far-out ideas that will probably always remain so. It is instructive to note that Lida Mattman's introductory historical chapter on bacterial variation in her textbook *Cell-Wall Deficient Forms* does not mention Béchamp. She begins instead with the Swedish biologist Ernst Almquist, whose contributions on bacterial variation did not presume such radical pleomorphism.

Having thus situated Béchamp without dismissing his place in the history of microbiology, let us consider now the evaluation of the North American research tradition on bacteria and cancer. My evaluation will be organized around topics derived from the four major groups of theory-choice criteria. Because the consistency arguments provide a first line of evaluation, and because the materials included under these arguments provide helpful background information, they are discussed first. I have divided the consistency arguments into two groups based on background research in the microbiology of cell-wall deficient (CWD) bacteria and in the molecular biology and immunology of cancer.

The Consistency Argument I: CWD Bacteria Research

To evaluate the possibility that bacteria may be etiological agents in cancer, it is helpful to begin with the status of research on bacterial variation or pleomorphism. By the late 1940s and 1950s the cyclogenist controversy was largely resolved against cyclogeny, at least in the strict forms proposed by Enderlein and Hadley, but the filtrationist controversy was resolved in favor of filtration.[4] In a chapter on filterable forms, Mattman writes, "Demonstrating filterable life in microbial species frustrated some investigators, was accomplished by many pioneer microbiologists, and now can be a classroom exercise" (1993: 209). Furthermore, even at the height of these controversies during the 1920s and 1930s, the general phenomena of variation were accepted, and instances of bacterial variation are easy to see under the microscope. By 1960 the topic of bacterial variation was well-enough accepted that a major multivolume collection on bacteria included an essay on the topic (Klieneberger-Nobel 1960).

The phenomenon of bacterial pleomorphism is so widely accepted in microbiology today, at least for certain groups of bacteria, that it can be considered consensus knowledge. For example, a major microbiology textbook mentions bacteria pleomorphism as a common-knowledge, taken-for-granted phenomenon that needs no citation (Atlas 1988). One needs only to check Medline or another common index under CWD bacteria or L-forms (a plasmalike, fried-egg colony appearance) to see that there is continuing research on the topic in the major peer-reviewed journals.

Nevertheless, there is still some stigma attached to the topic. In a conversation I had with Gerald Domingue, one of the foremost experts in the field, I learned that although the existence of L-Forms and CWD bacteria is not controversial among microbiologists, the theory that they can persist in the host and play a role in pathology is still controversial. As he notes in the volume *Cell Wall Deficient Bacteria: Basic Principles and Clinical Significance,* "These aberrant bacteria have been regarded as laboratory curiosities of little or no clinical significance" (1982: ix). That perspective seems to be changing, particularly with the publication of Domingue's book and the second edition of Mattman's book (1993), which carries the subtitle "Stealth Pathogens." Many medical experts believe that without cell walls bacteria that remain in the host body would simply rupture under osmotic pressure. However, Domingue and others argue that CWD forms of bacteria may survive by entering host cells (possibly integrating with host-cell organelles) where they may persist as buried bacterial genomes. The theory is bolstered by the widely accepted evolutionary theory that mitochondria were once captured bacteria that eventually evolved into necessary cell organelles.

Domingue's work has been central to documenting the role of buried bacterial genomes as cryptic agents of chronic disease. He and associates documented for the first time in 1974 that a relatively stable L-form was definitely linked to the phenomena of persistence and reversion in a human embryonic tissue culture system. Small, dense bodies derived from an unstable L-form in the tissue-culture system were observed to persist for a prolonged time, after which the bodies reverted to the classical parent organism. This was accompanied by death of the tissue-culture cells. As a result of these studies Domingue argued that the bacteria have a reproductive cycle in which undifferentiated dense forms ("stem cells") are extracted from the vesicles of the "mother" L-form along with elementary bodies. (Thus, the concept of a bacterial cycle reenters the literature, but in a much different way from the cyclogenic theories of the 1920s.) Domingue argues that the undifferentiated dense forms increase in size and are capable of maturing into the vesiculated mother forms as well as reproducing or reverting to the bacterial form. His current research (especially for idiopathic hematuria) supports his hypothesis that intracytoplasmic dense bodies are a mechanism for bacterial persistence and may cause infections systematically overlooked in clinical medicine. He adds, "Persisting small, electron dense, elementary bodies derived from bacteria with aberrant cell walls seem to have cell and tissue tropisms correlated with a variety of chronic human diseases, notably chronic infections of the urinary tract."[5]

Domingue argues that under conditions of immune deficiency or trauma, CWD bacteria may regain their standard morphology and their pathogenicity. His colleagues sometimes tease him about his research on "funny bugs," but they also send him samples to analyze when they are having difficulties with diagnosis and treatment. In addition, he has received continued NIH funding for his work on CWD bacteria and their possible role in interstitial cystitis, and he has successfully treated many patients who have been afflicted by chronic urinary-tract diseases associated with CWD bacterial infections. There is growing interest in the recent research on CWD bacteria and disease, and Domingue regularly receives invitations to speak before groups of scientists and doctors on this increasingly hot topic. The recent research on CWD bacteria can help make diagnosis and culturing more sensitive to possible bacterial variation that may have been missed under standard procedures. Diagnosis of CWD bacteria may also lead to better selection of antibiotics, such as ones that operate through pathways other than destroying cell walls.

As stealth pathogens, CWD bacteria have been implicated in a number of chronic, degenerative diseases. Mattman's review (1993) of these diseases includes urinary-tract infections, Crohn's disease, leprosy, meningitis, multiple sclerosis, myocarditis, rheumatic fever, sarcoidosis, septicemia, and ulcerative colitis. Probably the most well-known case is arthritis and other collagen diseases, where there is a long tradition of researchers who have advocated a bacterial etiology (Hughes 1994). In the United States Thomas Brown was for many years a relatively isolated medical doctor who supported the use of antibiotics in the treatment of arthritis. However, by the 1990s the field was in the midst of a major controversy as randomized clinical trials of monocycline indicated a statistically significant improvement over placebo controls.[6]

What do the leading CWD researchers think about bacteria as agents of possible etiological significance in cancer? Domingue, Mattman, and other researchers in this area are careful not to extend the claims of research on pathology and CWD bacteria to cancer. Based on my conversation with Domingue, I would characterize his attitude as open-minded and intrigued but skeptical. Mattman includes a chapter on the topic in her textbook, but she also cautions, "There is no subject generally viewed with greater skepticism than an association between bacteria and human cancer." She then adds,

> However, the medical profession may look back with irony at the stony reception given by his colleagues to Koch's paper elucidating the etiology of tuberculosis. Similarly, medical students were once taught that whooping cough vaccination was an unrealistic dream reported only by two women at the Michigan Public Health Laboratories and by a deranged pediatrician named Sauer. (1993: 318)

The question now seems to be not whether CWD bacteria play some role in at least some cases of chronic, degenerative disease, but how important that role is. The experts on CWD bacteria are not willing to exclude the possibility that cancer may be among the diseases for which bacteria play some role.

The case of ulcers is one example of a huge shift in medical opinion involving the reexamination of bacteria and chronic, degenerative disease. Long considered a noninfectious disease caused by stress and lifestyle factors, the maverick research of the Australian doctor Barry Marshall led to a consensus shift in medical opinion during the late 1980s and early 1990s (Marshall et al. 1988; Marshall 1994). In 1994, the National Institutes of Health held a consensus conference that resulted in substantial changes in the recommended treatment for ulcers in recognition of the new theory.

During the 1990s an increasing number of research reports linked *Helicobacter pylori* infection to stomach cancers (e.g., Eurogast 1993; Parsonnet et al. 1994). Here is one case, then, of a relatively noncontroversial linkage of bacterial infections to cancer etiology.

The reevaluation of ulcers, not to mention the potential changes in medical opinion on the etiology of arthritis and other chronic diseases, suggests that research on bacteria and cancer *could* become part of a general reevaluation of the role of bacteria, particularly CWD forms, in chronic disease. In other words, there is no case for prima facie dismissal of the bacterial theory as inconsistent with expert opinion on CWD bacteria and pathology. The more one reads in the area of biomedical research, the more one realizes how little we all know and how foolish are those who dismiss reasonably grounded medical hypotheses out of hand.

Consistency Argument II: Cancer Research

The second area of background knowledge that provides a reference point for evaluation by consistency is contemporary cancer research. A relatively noncontroversial area of contemporary cancer research involves the use of bacteria or bacterial products to stimulate the immune system (e.g., Jeljaszewicz, Pulverer, and Roszkowski 1982). The most well-known products are Coley's toxins and the tuberculosis vaccine BCG. This area of research is relatively noncontroversial because it is not guided by the theory that bacteria may be etiological agents in cancer. Mechanisms proposed to explain the efficacy of bacterial vaccines involve theories such as the cascade of cytokines that the bacterial vaccines are presumed to release (e.g., Wiemann and Starnes 1994). These mechanisms locate the use of bacterial vaccines within contemporary immunology and molecular biology.

As in most other areas of biomedical research, molecular biology has become the obligatory point of passage for the bulk of cancer research, and since the 1970s a large number of researchers has climbed aboard the proto-oncogene theory bandwagon (Fujimura 1995). Under this theory, cancer emerges when proto-oncogenes are activated to form oncogenes, or conversely when tumor suppressor genes become inactivated. The proto-oncogenes regulate the expression of growth factors and growth-factor receptors. When they are not functioning properly, the cell's growth regulation mechanisms are thrown off. Cancer researcher Robert Weinberg describes the current understanding of this process with an automotive metaphor: "In effect, a cancer cell's growth may derive from a stuck accelerator (an activated oncogene) or a defective braking system (inactivated growth-suppressor gene)" (1994: 166). The genetic changes that lead to cancer generally occur gradually over time; thus, cancer is a multistage process that can begin with inherited genetic defects that provide a predisposition and that may require additional, subsequent genetic damage from carcinogens.[7] The molecular biology of cancer therefore highlights multistage genetic damage, as well as protein and growth-factor expression often at the cell walls. That emphasis provides a point of connection among molecular biology and related sciences such as immunology and endocrinology. This theory of carcinogenesis is sometimes called the mutagenic theory; however, I prefer the term "molecular" because the contemporary understanding of tumor genesis and growth includes other phenomena such as the failure of gene-repair mechanisms.

In contrast to contemporary molecular approaches to cancer, research on bacterial pathogens appears to be out of another epoch, the bacteriology of the modernist period rather than the molecular biology of research today. The morphological nature of most of the bacteria-and-cancer studies prior to the 1960s is in striking contrast to the molecular nature of much cancer research today. The two research traditions, one great and the other little, speak different languages, use different technologies, and, to invoke Kuhn's term, appear to operate under incommensurable paradigms (Kuhn 1970). This appearance of incommensurability provides reasonable grounds for the argument that there is general inconsistency between the bacterial-etiology theory and the bulk of contemporary cancer research. Therefore, on first inspection the theory that bacteria are carcinogenic agents could be rejected for inconsistency with contemporary cancer research.

However, the inconsistency is not necessarily as great as it first appears. Almost all of the contemporary research on the molecular biology of cancer began with the study of oncogenes or proto-oncogenes that were first associated with tumor viruses. In the process, viral oncology became an integrated part of modern human-cancer research (Fujimura 1995; Rettig 1977). For bacterial research to accomplish a similar transition, it would have to be translated into the language of molecular biology, immunology, endocrinology, and related fields that constitute the core of contemporary cancer research. This translation problem is not restricted to the bacterial theory; instead, it is a general problem among many of the alternative cancer theories and therapies. Most of the so-called metabolic researchers are concerned with nutritional aspects of cellular metabolism, and they tend to think in terms of the biochemistry of metabolism rather than the molecular biology of gene expression or gene repair. The failure to make the necessary translations into the language of molecular biology and related fields therefore constitutes a major hurdle to acceptability on the part of many alternative cancer theories and therapies.

Notwithstanding the apparent incommensurability, one might recall the lesson of STS analysts Barry Barnes and Donald MacKenzie, whose critique of Kuhn's paradigm-conversion theory pointed to the empirical problem with his theory: opponents from clashing paradigms or research traditions often can become quite adept at understanding the terms of the other side (Barnes and MacKenzie 1979: 200). Although the molecular researchers and the metabolic/nutritional researchers may not speak each others' languages, translation is possible in principle. In other words, the two fields are not contradictory or incommensurable as much as operating at different levels.

Mattman reviews a number of possible means by which bacteria could play a role in tumor genesis and growth from the perspective of contemporary immunological and molecular theories of cancer (1993: 314–16). One set of possibilities involves blocking host immune response. Viral models include the Friend leukemia virus, which prevents normal antibody response, and the Gross leukemia virus, which leaves antibody formation intact but prevents cellular defense. For bacteria the major candidate at present is their production of a substance similar to the human hormone choriogonadotropin, which is found along the cell walls of cancer cells and may block immune response. The viral models of blocking immune response have no bacterial research yet to support them, but the choriogonadotropin theory has been researched in some detail for bacteria and will be discussed shortly.

Mattman also provides two other possibilities that could involve a relationship between bacteria and oncoviruses. First, some studies suggest that viruses and bacteria together may be necessary to produce a murine leukemia. Second, bacteria may host pathological oncoviruses, following the model of *Aerobacter aerogenes* and Influenza-A virus, which the bacterium holds on its surface (Mattman 1993: 314–16). A similar mechanism is influence on the host DNA via plasmid transfers, a mechanism that was first suggested for the plant pathogen *Agrobacterium tumefaciens* (Macomber 1990).

Another possibility is that bacteria may produce a carcinogenic compound known to damage host DNA. The most well-known case is aflatoxin, a carcinogen produced by the plant fungus *Aspergillus flavus* and known to contaminate human foods such as peanuts. There is also some evidence for this kind of mechanism in murine models (Laquer, McDaniel, and Matsumoto 1967).

Although future research could prove that some or all of these possibilities are dead ends, the list is sufficient to meet the incommensurability or nontranslatability argument. In other words, it is possible to make the theory of bacteria as carcinogenic agents consistent with the contemporary molecular biology, immunology, and/or endocrinology of cancer. However, a second problem immediately faces the theory on consistency grounds. Under the contemporary molecular research program, the etiology of cancer is understood as a multifaceted process that involves a wide range of possible carcinogens. To bring bacteria into the picture, one might relegate them to minor factors in a few cancers, such as *Helicobacter pylori* in gastric cancer. This strategy, similar to the role of viral oncology, would be the easiest way to transform the bacterial theory into something acceptable and useful to contemporary cancer research. However, the strategy would involve a very different understanding of the role of bacteria in cancer from that envisioned by most of the researchers, and thus before endorsing that strategy it is necessary first to consider and reject the older, more comprehensive bacterial theory.

Under the molecular theory, the direct cause of cancer (as opposed to indirect variables such as lifestyle or poverty) is any agent that can bring about genetic changes to alter sufficiently the growth signalling mechanisms of the cell. Under the subtheory of multistep carcinogenesis, genetic transformation from a normal cell to a cancer cell is assumed to involve a mixture of inherited predisposition and environmentally induced genetic hits that accumulate over an organism's lifetime and lead gradually to carcinogenesis. Recognized agents that are direct causes of mutations include radiation, chemical carcinogens, oncoviruses, and inherited defects. In addition to causing direct genetic damage, risk factors can contribute to tumor-promotion mechanisms and to the failure of gene-repair mechanisms or immune-system surveillance. Lifestyle factors such as poor diet and (in women) number of estrogen cycles may fit more into this category.

Some of the early researchers discussed in the second chapter developed a view of carcinogenesis that appears to be very inconsistent with the contemporary theory that radiation and chemical carcinogens cause cancer directly. Recall that Rife and Crane argued that the dead tissue left by X-rays formed a "natural parasitic feast." Glover made a similar but much more developed argument regarding the Japanese coal-tar studies:

> The outstanding facts in these tar experiments are:
> 1) The irritation must be continued for a long time before cancer develops.
> 2) Only a small percentage of the experimental animals ever develop cancerous tumors, while the majority show inflammatory or papillomatous lesions.
> 3) The carcinomata *[sic]* always start in small isolated areas and never in the whole of the irritated areas.
> 4) Most of these irritated areas have continuous open wounds which render them vulnerable to infection.
>
> It has been argued that some chemical substance in the irritant acts on certain small groups of cells in the stimulated area, changing them into proliferating cells and that the proliferative power of the cell itself is the cause of the malignancy. This does not seem plausible in view of the fact that Rous Sarcoma No. 1, which is a true malignant tumor, can be transmitted by an inoculation of dead cells or a cell-free filtrate. (1926: 165)

In short, Glover and Rife suggest that radiation and chemical carcinogens do not directly cause cancer; rather, they set the stage for the action of a microbial pathogen.

One can find more or less the same view among some of the tumor virologists during the period prior to the molecular biology of cancer, such as Peyton Rous (1941) and Francisco Duran-Reynals (1950). Rous provides perhaps the clearest statement of the theory that chemical carcinogens activate latent viruses:

> Experiments with carcinogenic chemicals have disclosed the fact that every mammal carries within its tissues thousands of potentialities for tumor formation, and yet no growth results from any of them unless the circumstances are peculiarly favorable. If viruses furnish the potentialities mentioned, they may enter the body in infancy, as do bacteria, and be distributed to the tissues, persisting there as harmless inhabitants associated with cells until altered circumstances render them injurious. . . . The "carcinogenic agents," so called, may be thought of as providing the conditions required for some innocuous resident viruses to alter and become tumor-producing viruses. (1941: 39–40)

The views of Glover, Rife, and tumor virologists such as Rous and Duran-Reynals are no longer consistent with current knowledge. Radiation and chemical carcinogens appear to cause genetic damage that leads directly to carcinogenesis without necessary mediation from microbial agents. Thus, from the point of view of today's molecular theory of cancer,

Glover's argument that "irritation must be continued for a long time" could be met by explaining that the long time lapse was the result of the need for a multistage sequence of genetic damage or a gradual failure of gene-repair mechanisms through constant stress and irritation. Furthermore, his argument that only a small percentage of animals develop tumors could be explained either because chemical carcinogenesis usually involves an initiator and a promoter or because the laboratory animals used at the time were not all genetically similar. Although Glover uses the argument of cancer foci as evidence for his theory, the opposite is the case today, when the evidence now is in favor of monoclonal tumors (that is, tumors originating from one cell rather than many). This finding is generally interpreted as evidence in favor of a noninfectious etiology (e.g., Varmus and Weinberg 1993: ch. 2). Thus, it is easy to answer Glover's and Rife's arguments from the perspective of the secondary-infection theory nested within a molecular theory.

However, rather than reject Glover and Rous out of hand, let us give them the benefit of the doubt and imagine how they and other researchers might respond today by drawing on the current molecular understanding of cancer. As in the case of the herpes virus, they are envisioning a latent cancer microbe that is the result of an original infection, perhaps even passed on through "vertical transmission" at birth (Gross 1983). The microbe is probably an intracellular pathogen or, if extracellular, it is inaccessible to the blood system and immunological surveillance. If intracellular, the microbe probably lacks cell walls and may be the size of a virus. The microbe remains in a latency phase until conditions of cellular dysfunction caused by (in this case) radiation or chemical irritants cause the microbe to emerge from a latency phase. Smoking, poor diet, and other known risk factors are other possible triggers for a transformation from the latency phase. Once activated, the microbe then contributes to transforming the host cell into a cancer cell through one or more of the mechanisms outlined above. If the virus or bacterium is a latent, intracellular pathogen, it is possible that many tumors would be monoclonal in origin. In short, if we translate the bacterial theory a little, it is possible to reconcile it with today's theory of multistage carcinogenesis.

Let us call this the "modified bacterial-infection theory." As for the contemporary viral theory of cancer, bacteria need not be present for all cancers, but they may be included as contributors to the multistep process of genetic damage and failed gene-repair mechanisms that are assumed today to be the molecular mechanism for tumor genesis. This modified bacterial-infection theory would pass the test of consistency with respect to the contemporary molecular understanding of cancer. In other words, a modified bacterial-etiology theory could replace the secondary infection theory. Latent bacterial pathogens therefore could serve as contributors to some steps in the multistage carcinogenesis of at least some cancers.

At this point an advocate of the secondary-infection theory could raise another consistency argument: if bacteria were involved in cancer, the use of antibiotics in cancer patients would have been efficacious. There are a couple of ways to answer this argument. First, if the bacteria are intracellular pathogens, or if their location is outside the host cells but in positions not accessed by the bloodstream, it may be difficult for antibiotics to reach them. Second, because the bacteria have a long-term relationship with the host, they may have developed strains that are resistant to antibiotics (Gregory 1952: 132). It is now well-known that some forms of common bacteria have developed strains that are resistant to all antibiotics; it is certainly possible that possible oncobacteria have developed antibiotic-resistant strains as well.

However, this argument would suggest that when antibiotics were introduced, their use should have led to the successful treatment of cancer. I know of little evidence to support this claim except the work of one American doctor (Gregory 1952). Thus, an alternative defense would be better: the pleomorphism of the proposed oncobacteria and their intracellular location may provide them with a defense against antibiotics. On this point it is worth noting that CWD forms can be induced experimentally by some antibiotics, such as penicillin, which works to break down cell walls. (This fact raises the possibility that some of the increase in cancer rates could be due to the use of antibiotics that destroyed bacterial cell walls but did not kill the bacteria.) In summary, by assuming difficulty of access or resistance to antibiotics (either through the development of resistant strains or through pleomorphism), it would be possible to make the claimed lack of efficacy of antibiotics consistent with contemporary medical knowledge.

Probably a better rebuttal to the antibiotic argument is to note that some antibiotics in fact *are* efficacious in the treatment of some cancers. Contemporary chemotherapies such as the anthracyclines are efficacious for some cancers, particularly Wilms's tumor, leukemias, and lymphomas (Priebe 1995). Of course, the mechanism is presumed not to involve bacterial infections; instead, cancer cells are just slightly less resistant than healthy cells to the toxicity of the drugs. Nevertheless, the antibiotic treatment of *Helicobacter pylori* appears to lead to regression of gastric lymphomas (Wotherspoon et al. 1994). A supporter of the bacterial theory might argue that the success of antibiotics in some cases may be due in part to an effect on bacterial pathogens. Furthermore, if cancer-associated bacteria are highly pleomorphic, then it is possible that the best antibiotic choices have not yet been tested. For example, Domingue (1996) is careful to point out the importance of appropriate antibiotic selection (nitrofurantoin) in his successful treatment a case of hematuria for which he cultured CWD bacteria from catheterized samples.

In conclusion, an advocate of bacteria as carcinogenic agents can meet the consistency arguments that are raised against it, but only if the theory is modified significantly so that bacteria are viewed as contributing agents to multistage carcinogenesis. At its best the modified bacterial theory could provide some new ideas for treatment in conventional cancer chemotherapy, such as specific new antibiotics, as well as possibly new protocols involving long-term treatment and mechanisms for appropriate absorption. The consistency argument from antibiotics may then result in what Michel Callon (1986) calls problematization, or provoking others to become interested in a new or different theory. By showing how current research on antibiotics and cancer might obtain new successes when considered from the viewpoint of the bacterial-infection theory, the antibiotic counterargument can be turned on its head and used in favor of the theory. Politically, of course, this interpretation also repositions the theory so that it is no longer opposed to the powerful pharmaceutical industry. Rather, the theory maintains a role for antibiotics alongside vaccines; it may even provide possibilities for new and more efficacious uses of antibiotics in the treatment of cancer.

The bacterial theory can also be aligned with the alternative (or adjunctive) nutritional therapies. For example, Livingston argued that abscisic acid, a plant hormone and Vitamin A analog, neutralizes the production of substances similar to the hormone choriogonadotropin, which she argued is probably the major pathway by which bacterial infections contribute to cancer (Livingston 1979). Thus, the dietary part of her therapy became aligned with the vegetarian diets that are common among the metabolic cancer treatment protocols such as the Gerson therapy, which uses raw vegetable juices, including that of carrots. However, Livingston still viewed the bacterial vaccine as the key element. In a subsequent paper she measured percentage inhibition on murine sarcoma 180 in vivo and found 20 percent for Vitamin A, 25 percent for abscisic acid, 70 percent for her purified antigen, 58 percent for abscisic acid and Vitamin A, and 76 percent for all three combined. She concluded that although the dietary supplements were helpful, her bacterial antigen was much more important (Livingston and Majnarich 1986). In a talk before the Cancer Control Society, she explicitly stated that while she thought the Gerson therapy was important, it was necessary to have the additional contribution of her vaccines (Livingston 1989). In any case, Livingston certainly opened the door to a linkage between nutritional therapy and the bacterial program.

The Accuracy Argument I: Contaminants

At this point the supporter of the secondary-infection theory could raise a different type of argument: even if a modified form of the bacterial-etiology theory can be made consistent with contemporary knowledge in microbiology and cancer research, there is no evidence to support the theory. The bacterial cultures claimed to be associated with human tissues in these studies are only artifacts of contamination from sloppy laboratory procedures. This counterargument shifts us from general consistency arguments to a group of accuracy or evidence arguments. I will break the accuracy arguments down into two separate groups: first, the laboratory techniques used to culture bacteria from cancer tissues introduce environmental contaminants; and second, there is no evidence that supports the existence of a single pleomorphic cancer organism.

The contaminant argument could begin with failures to culture bacteria from cancer cells. One of the main problems with the bacterial theory during the crucial period of its rejection at the beginning of the twentieth century was that some researchers failed to isolate bacteria from their tumor samples. Even today, there are cases of researchers who have attempted to replicate the work and have failed to culture bacteria from their malignant tissue samples. For example, a biochemist whom I met at a conference told me that he put a microbiologist on the project using the Livingston and Alexander-Jackson techniques, and the microbiologist was unable to culture the bacteria. Likewise, Mattman also failed to culture a purported cancer organism in her laboratory. In her words,

> Joseph Merline, in our laboratory, cultured over two hundred bloods from lymphoma patients and in no instance found the bacterium of Glover, Nuzum, or of more modern descriptions. Merline was not seeking a tumor-instigating bacterium, and such procedures as washing the red cells to elimi-

> nate antibody or aging the blood at room temperature to void complement were not followed. His study rather shows that careful standard technique to isolate aerobes, anaerobes, and CWD forms does not grow a carcinogen-fostering bacterium. (1993: 316)

The problem with negative instances is that defenders of the bacterial theory can always argue that a failure to replicate is due to a failure to use the appropriate methods. Mattman leaves room for this interpretation in her description of her laboratory's failure to replicate. She adds that some of the difficulty to replicate is because much of the media used is exotic, such as "Glover's concoction of sunflower seeds, Iceland moss, and Irish moss" (316). We are therefore in the situation first delineated in science studies by Harry Collins (1985) as the "experimenter's regress": defenders of the bacterial-infection theory can always argue that a failure to replicate means that the experimenters made a mistake in their procedures.

Conversely, attempts to refute the technique of culturing bacteria from mammalian blood or tissue samples suffer from the same problem of the experimenter's regress. For example, two critics first obtained a pleomorphic, acid-fast organism from the blood of Hodgkin's disease patients, but later, when they used similar procedures and cultures in more aseptic conditions and under a sterile hood, they were unable to culture the organism (Kassel and Rottino 1955). They suggested that their failure to culture the bacteria under more rigorous conditions provided the definitive crucial experiment that refuted the bacterial theory. However, a defender of the theory can always argue that Robert Kassel and Antonio Rottino introduced something into their second protocol that caused the failure to replicate. One is only limited by creativity: the failure to replicate could be explained by Chlorox on the gloves, the aerosol spray used to clean the inside of the chamber, temperature differences, oxygen differences, light differences, and so on. At best, then, the experiment provides evidence in favor of the contamination interpretation—it is one piece in the puzzle—but overall evaluation will always be a much messier process. Once again, evaluation of major theories in science is not reducible to an algorithm; it is much more like arriving at a verdict after having heard various witnesses in a court trial. Kassel and Rottino are two more witnesses for the prosecution.

The critics probably would have been more convincing, and had an easier time, if they had adopted another tactic: to find some loophole in an experimental design and argue that an artifact was introduced because of this design. Critics therefore have a politically advantaged position because they only have to read the opponents' experiments and come up with possible design flaws. In contrast, to meet the critics' arguments, advocates need to reanalyze or redo experiments, which is much more costly in terms of time and resources. This tactic can be deadly for out-groups that do not have access to mainstream financial resources. Let us adopt this position for a moment in favor of the secondary infection theory, and examine the best case experiments for evidence of possible artifact by contamination.

Glover was aware of the contaminant argument, and from comments made in his 1938 report he probably had heard many versions of it. One should recall the political context of the 1938 report, in which there was opposition to his research within and outside the National Institute of Health. Many people were probably looking over his shoulder with the hope of finding artifacts, and in his 1938 report Glover was careful to head off their potential criticisms. He boiled his media, passed it through a bacterial filter, and used sterile equipment in its manufacture. He also incubated all media to see if it produced any microorganisms that might have been latent in the media, and he presented a table that showed that all tests were negative. He used control slides to protect against a staining artifact, and he inoculated control flasks with sterile saline to make sure that artifacts were not introduced from injection. In his animal experiments he used control tissues and unplanted media, which remained sterile. In the 1938 report, he also describes incubating the organism from a malignant breast tissue that was received in a sealed sterile container fresh from the operating room. A description of the technique used in the 1930 report gives an additional indication of his concern with sterility:

> In a "sterile" room and under aseptic conditions, the skin is reflected, the surface of the abdomen seared and opened. By means of a sterile loop specimens are taken from the peritoneal cavity and rubbed over the surface of the slant medium. The chest is then opened, the heart seared, blood aspirated from its chambers and transferred to tubes of the slant medium. (1930: 102–3)

One might argue that Glover is a weak witness for the defense, because of his dishonesty regarding the issue of credit for his work. However, no one claimed that his research per se was fraudulent. Furthermore, other researchers such as Alexander-Jackson have described similar precautions in their protocols. Her measures included boiling, filtering, and autoclaving (even twice) the medium. When she removed blood, she cleaned the skin with iodine and alcohol, and expelled the first drops of blood prior to expelling the rest into the broth (1954: 40–41). She used controls from blood-bank donors and healthy, young, tuberculin-negative nurses, and they were free of the organism. (She did not like to use controls from the general population, because according to a personal communication she had from Glover, he believed that as many as 20 percent of normals and 40 percent of the elderly tested positive for the cancer-associated bacteria.)

A second argument against contaminants is that many of the studies showed partial or full completion of Koch's postulates. Partial completion involved isolating the organism, injecting it into the host animal, and then analyzing the animal for signs of malignancy. This was achieved as early as the Nuzum studies in the 1920s, but also in William Crofton's work and in studies in the 1960s and 1970s, such as those by Sakea Inoue of an acid-fast microbe associated with tumors in the newt.[8] Diller also claimed that over many experiments she was roughly doubling tumor production in one strain of mice that were injected with a microorganism isolated from rat sarcoma when controlled against noninjected matched mice (Diller 1974; Diller, Donnelly, and Fisher 1967). However, the Inoue and Diller studies do not count as complete fulfillment of Koch's postulates, because they did not report on reculturing the organism from infected animals.

By far the most complete and meticulous report of the fulfillment of Koch's postulates was again Glover's 1938 report. He cultured bacteria from the animal over six successive subcultures to reduce the possibility that the bacteria were contaminated with any substance from the tissues of the organism. He then reinjected the animals (guinea pigs because of their low natural incidence of cancer) and repeated the entire procedure over six "generations" or repetitions, each time with different sets of animals. It is important to review the Glover studies because misinformation continues to be disseminated today. The American Cancer Society report on Livingston's research appears to be at least partially erroneous when it states, "Investigators at the National Institutes of Health, however, could neither substantiate the presence of a microbe nor duplicate the vaccine from organisms supplied by Glover" (American Cancer Society 1990: 104). Many years later, one researcher appointed by the government reported that he replicated Glover's work by reculturing the bacteria from dead guinea pigs (Clark 1953). Alexander-Jackson also wrote that she fulfilled Koch's postulates completely and successively by "repeated reisolation of infected animals of cultures of the inoculated organism" (1954: 47).

A more roundabout version of the contaminant or artifact argument would be to grant the case that the researchers were inducing a disease through bacteria, but then to deny that the disease was true cancer. This appeared to be at the heart of the scientific justification for stalling publication of Glover's work in the 1930s. In the preface to his 1938 pamphlet, medical doctor George McCoy, the former director of what became the NIH, outlined the controversy among the histologists regarding the interpretation of the guinea-pig "malignancies." Because cancer is rare in guinea pigs, histological proof of malignancy from inoculations with the Glover organism would have strengthened his argument that he had indeed found the cancer microbe. It appears that the controversy among the pathologists was never resolved for the guinea pigs. However, McCoy adds that they soon turned to rats:

> As was expected, tumor production was more readily secured in the rat and there was little or no difference of opinion among the pathologists as to the nature of the gross and microscopic appearances. It is to be noted, however, that in some experiments lesions of uncertain nature in the guinea pig, and of unquestioned malignancy in the rat, were apparently induced by the same agent; this of course suggests that there may be at hand a method that will throw light on the pathogenesis of new growths. (Glover and Engle 1938)

McCoy also states that he could not rule out the possibility that the malignancy was caused by a virus that accompanied the bacterial cultures or that some component of the media was carcinogenic. The virus explanation, however, is a version of the infectious theory, and if the virus turned out to be a filterable stage of a bacterium or a bacteriophage, it would mean returning to the bacterial theory. Thus, from the perspective of the secondary-infection theory, it is a weak counterargument. The carcinogenic-media argument was a good one for the time, but since then many different media have been used, and they have varied significantly, been autoclaved, and cultured as controls to demonstrate sterility. Furthermore, the media-carcinogen argument would not explain what Glover saw as central evidence against the contamination argument: the success of his serum. This claim for efficacy of the serum—or vaccines in the Livingston and Seibert research—shades into the pragmatism criteria and will be discussed shortly.

In conclusion, although there is no way to achieve closure against the contamination argument, the standard of a "reasonable scientist's design" might be invoked, much as the standard of the reasonable person is invoked in legal disputes. The fulfillment of Koch's postulates is usually considered a reasonable standard in microbiology, especially if there are controls and if vaccines or sera protect against challenges with the pathogen. My evaluation, then, is that the bacterial-infection theory withstands the accuracy counterargument in the form of artifact by contamination. Of course, it would be better to have contemporary studies using contemporary methods and controls, but given the available evidence, the argument for contamination is not strong enough at this point to justify rejecting the theory. However, the theory does not pass the second accuracy argument, and in order to survive, it will have to be modified a second time.

The Accuracy Argument II: No Single Cancer Organism

The second accuracy argument relies on inconsistencies among the advocates. For example, Glover, Enderlein, and Naessens organize the stages into a single cycle, whereas Livingston, Alexander-Jackson, Diller, Seibert, and other researchers affiliated with that group do not. This returns us to the cyclogenist controversy. Because the consensus among microbiologists today—even among those who study CWD bacteria and its filterable forms—is against the strong form of cyclogeny that was popular during the 1920s and 1930s, it makes sense to reject the cyclogenic theories of Glover, Enderlein, and Naessens and accept instead the multistage, noncyclogenic phases described by Livingston and colleagues. Furthermore, the cyclogenist position can also be explained as the product of modernist cultural thinking that tended to oversimplify natural and social processes into neat and closed systems based on equilibrium principles (Hess 1995: ch. 4). In contrast, the Livingston network admitted transitions among stages but did not order them into a neat cycle. This position would also correspond to contemporary CWD bacteria researchers such as Domingue, who has outlined a modified or limited cyclical model of pleomorphism.[9]

A more difficult internal contradiction is that the observed stages appear not to coincide across researchers. For example, Gerlach believed that his "parasitic fungus" was not the same as the von Brehmer organism (in Diller 1962a: 203). Glover postulated a fourteen-stage cycle, but the stages do not coincide, at least not in any obvious way, with those of the sixteen-stage cycle of Naessens or the fourteen-stage cycle of Enderlein. The Enderlein cycle includes more submicroscopic or filterable forms than either the Naessens or Glover cycle, and the Naessens and Enderlein cycles include funguslike phases that appear to be missing in the Glover cycle. However, many of the middle stages of the Enderlein and Naessens cycles are comparable to the Glover cycle. This could be interpreted in two ways: Glover only observed part of the cycle, or Enderlein and Naessens were observing multiple microorganisms, such as a fungus and a pleomorphic bacterium similar to the mycobacteria, coryneform group, or actinomycetes. I am inclined to the view that Enderlein and Naessens were observing more than one microbial species. This interpretation is consistent with the subsequent observations of the Livingston network, whose observations of pleomorphism were more restricted and did not include large fungal stages.

Thus, to defend the bacterial infection theory against the argument that the observations were not accurate because they were inconsistent with each other, one has to side with some researchers over others. I suggest accepting the post–World War II North American group as the most credible. In general, North American science after World War II was the best in the world, and this research group was interacting with colleagues in this scientific community and publishing in their journals. They also had the advantage of reading the earlier research of Enderlein and Glover. In contrast, the other researchers were either historically prior to this group or not publishing in scientific and medical journals.

What, then, did the organism look like according to this group? In 1954 Alexander-Jackson presented a summary of the forms observed in the proposed cancer organism. The forms observed in tissue samples included submicroscopic bodies of 20–70 mμ revealed by electron microscopy; intracellular, Seitz-filterable, acid-fast bodies of 0.2μ; "larger coccoidal forms, often connected by slender filaments, [that] range in size from 1.5μ to cystlike ring forms 3–4μ in diameter"; and larger "globi" reminiscent of leprosy that suggest fungal budding forms and can be confused with the host cell nucleus if not differentially stained (Alexander-Jackson 1954: 45; also Livingston et al. 1956). When cultured in their media, the microorganism usually took a form that she described as a "pleuropneumonialike zoogleal matrix or symplasm" (like a mycoplasma or L-form) with coccoid forms that lengthen out into rods or into a cystlike ring. The outlines of Alexander-Jackson's observations in the 1950s are maintained in an article published by Livingston and Alexander-Jackson in 1970, which summarized the purported microbe as having seven major phases: virus-size, elementary bodies of 0.2μ, coccoid forms, ring forms, symplasm, rods, and filaments/spores in rods.

Subsequent research by Diller (1962b) confirmed "the essential features" of the stages outlined by Alexander-Jackson, but she added in another publication that the organisms she found were similar to those studied by Glover, Clark, von Brehmer, Villequez, Gerlach, Crofton, and Scott (1962a: 202). As I have demonstrated, this was only true up to a point, because only some of the stages corresponded across this group of researchers. Diller also dismissed the cyclogenic ordering of the various stages (1962a: 202). She did test isolates from von Brehmer and Villequez and found that they developed partial acid-fastness on appropriate media (203). In a subsequent paper she added that an "easily disrupted mycelium occurred infrequently in these cultures when the organisms were inoculated into von Brehmer's medium. Rod forms arose from the mycelium by fragmentation" (Diller and Medes 1964: 126). This comment suggests that some of the differences in observations may be attributed to different media, a conclusion that Rife and Gruner also drew from their comparisons.

A related accuracy argument switches levels and points to inconsistencies within the American group over classification. Livingston originally thought of the bacterium as a new species she denominated *Mycobacterium tumefaciens* because of its acid-fast staining property, and L'Esperance originally classified the organism as *Mycobacterium avians.*[10] This would place cancer in the same category as leprosy and tuberculosis, a classificatory decision that is strengthened by some of the gross similarities among the three diseases. However, subsequent research, particularly that of Diller, indicated that the bacterial samples stained acid-fast only in some phases or with some media. Because variable acid-fastness and the existence of motile forms are indications against classification as mycobacteria, the original classification came into doubt. Gram-positiveness, which Glover claimed to find consistently, is a characteristic of corynebacteria, and that possibility came under consideration. Diller found antigens in common with *Mycobacterium tuberculosis* and slight cross-reactivity with antiserum for *Corynebacterium diptheria,* and therefore she suggested an intermediate position between the two genera (Diller and Medes 1964: 128). However, Seibert introduced additional complications: "When cultural studies were made on two-day growths, some of our bacteria had most characteristics of *Staphylococcus epidermidis* and others of corynebacteria, but within these groups they were not identical, and all had some characteristics, such as motility and acid-fastness, that are not supposed to belong to these two prototypes" (Seibert et al. 1970: 725). Other researchers identified pleomorphic bacterial isolates from malignancies as *Staphylococcus epidermis* (Affronti, Grow, and Begell 1975) and *Propionibacterium acnes* (Cantwell and Kelso 1984).

In their publication in the *Annals of the New York Academy of Sciences,*

Livingston and Alexander-Jackson (1970) noted that sensitivity to bacteriophages and variable acid-fastness suggested classification as either mycobacteria or corynebacteria, but other criteria suggested similarities to listeria and the mycoplasma. They settled on the actinomycetales order, that is bacteria with branching filaments that resemble the fungi. Within that order they proposed a new family, "Progenitoraceae," because Livingston believed the bacteria may be archaic, and the genus "cryptocides," which means hidden killer. Probably Livingston and Alexander-Jackson chose the actinomycetales order because this group of bacteria is also known pharmaceutically for their use as producers of antibiotics, and they had isolated a toxic fraction from their cultures that they thought was related to the actinomycins (Livingston et al. 1970). The antibiotic Actinomycin-D is used in chemotherapy for Wilms's tumor, Ewing's sarcoma, and a few other rare cancers.

The impressive classification problems that I have outlined could be used to argue that the researchers were not working with the same organism. However, because pleomorphic bacteria are difficult to classify in general, the divergences do not necessarily provide evidence against the single organism interpretation. For example, another consensus fact passed along to students in a standard textbook of microbiology is the following:

> The mycobacteria and nocardioforms have traditionally been considered as related to the coryneforms and/or the filamentous actinomycetes, and the separation between mycobacteria, corynebacteria, and various pleomorphic bacteria is not easy. Different observers may classify the same strain as belonging to the genera corynebacterium, arthrobacter, nocardia, or mycobacterium. (Atlas 1988: 294)

Thus, the internal differences among the American group can, up to a point, be attributed to a general problem encountered among microbiologists.

The classification issue—and the single organism argument—is further complicated by the theory that the Rous sarcoma virus and perhaps other oncoviruses are filterable phases of larger microorganisms. Glover, Scott, and colleagues (1926) claimed to culture the Glover organism from a tissue sample of Rous chicken sarcoma No. 1 that Rous had supplied to them. They also claimed that they could culture their organism from malignant tissue infected with the Rous agent. Furthermore, Alexander-Jackson (1966) reported that she was able to culture, from the blood and tumors of Rous virus-infected chickens as well as from partially purified Rous virus, a pleomorphic organism with a cell-wall deficient phase. She claimed that injection of the bacterial isolate into chickens led to the development of the Rous disease in half the chickens but not in controls.

Alexander-Jackson (1970) also argued that even though the Rous sarcoma virus was classified as an RNA virus, DNA or DNA protein is present. She pointed to research that showed that the Rous sarcoma virus can be inhibited from multiplying and transforming normal cells by antagonists of DNA synthesis (Force and Stewart 1964). In addition, she cultured the virus in her broth and examined drops of the filtrate under an ultraviolet spectrogramic microscope, and she argued that the absorption peaks of the spectrograms indicated presence of DNA. She also presented electron microscopy evidence that suggested pleomorphism of the Rous virus. Controls of the broth alone showed no evidence for DNA.

I have classified Glover's and Alexander-Jackson's interpretation of the Rous virus as an accuracy argument because today the avian-sarcoma virus group is classified as an RNA retrovirus, and therefore one would reasonably suspect them of some kind of observational inaccuracy. However, Levinson et al. (1970) also found DNA in the Rous virus, and they concluded that the DNA was not due to absorption of cellular DNA. Although it is possible that the Rous virus may be a filterable phase of a larger microorganism, the one cultured by Glover and Alexander-Jackson may have also emerged from genetic recombination with host bacteria. In other words, the evidence appears to be ambiguous again.

The emergence of molecular biology has made it much easier to assess the theory that there is a unique, pleomorphic microbe that causes cancer. As Domingue explained to me, "With the new molecular technologies, people are forced to put up or shut up." In other words, it is possible to run molecular identifications of bacterial variants to determine whether or not they are species variants. He did this on two occasions with samples sent to him from the Livingston clinic and from a colleague at Rockefeller University. From his DNA-DNA hybridization analyses he concluded that some of the strains of "Progenitor cryptocides" were *Staphylococcus epidermidis, Streptococcus faecalis,* and *Streptococcus bovis* (Domingue 1982; Acevedo et al. 1987). Biochemist Hernan Acevedo also used standard morphological, nutritional, and biochemical measures to identify some of the Livingston "Progenitor cryptocides" strains, which he decided were *Staphylococcus epidermidis* and *Staphylococcus haemolyticus* (Acevedo et al. 1985). He also identified two strains from Seibert as *Staphylococcus epidermidis.*

In light of the studies of the 1980s by Domingue and Acevedo, one could continue to defend the theory that there is a specific, pleomorphic cancer organism by arguing that bacteria classified as ordinary staph and strep are in fact forms of highly pleomorphic organisms. However, this argument would be self-defeating because it would mean violating consistency criteria. There is one additional possibility. Note the warning issued by Domingue in his essay "Pleomorphic Cell Wall-Defective Bacteria as Cryptic Agents of Disease":

> The cultural techniques used conventionally by most clinical laboratories report only bacteria with cell walls and fail completely to recognize the large reservoir of cryptic organisms from which they derive in many chronic infections. An appreciation of the presence of these organisms could be achieved if cultural techniques appropriate to the growth of CWDB and methods to identify their DNA profile were used. Such studies would make an anachronism of *Bergey's Manual* as the great diversity within species became apparent, and would lead to a more meaningful appreciation of the etiology of many presently idiopathic diseases. It is interesting to speculate that ultimately classifications would tend to be less rigid as it is realized that microbes continually change as they adapt to various environments, changes that might also be found to include a shift of nucleotides between the host and other organisms, a kind of kaleidoscopic continuum with *in vivo* evolution of survival traits for microbes. (1996)

Domingue's comments recall Phillip Hadley (1927), although without the rigidity of his cyclogenic theory. Instead of the rigidity of the closed systems of the cyclogenic models of the 1920s, Domingue's description of bacterial variation suggests principles of open systems and flexibility that have become more common in scientific theories that have appeared after World War II (Hess 1995: ch. 4; Martin 1994). A more complete understanding of pleomorphic bacteria and cancer may have to await less rigid approaches to classification in microbiology. If these approaches were to materialize, the attempt to reduce the various stages to a single pleomorphic cancer microbe would be as wrong as the attempt to reduce them to known categories of standardized bacteria. Instead, one would need to work only with a more flexible framework of constantly shifting genomes. This is the direction that Domingue's work has pointed to since 1974, when he and colleagues documented electron-dense, buried bacterial genomes in embryonic cell cytoplasm (Green, Heidger, and Domingue 1974a, 1974b).

Let us, however, stay within the lines of traditional microbiological categories and assume that the DNA-DNA hybridization analyses mean that there is no single cancer organism. In other words, the researchers were viewing a variety of known organisms, some of which were pleomorphic. On the surface this appears to be a very strong argument against the bacterial-infection theory, and many scientists may close the book here and relegate the whole history to the premolecular era.

However, there is a surprising new fact that emerges out of the bacterial research program that gives the theory new life, even if one accepts the finding that there is no single cancer organism. In 1974 Livingston and her husband published a paper that detailed how under certain conditions their *Progenitor cryptocides* organism produced a substance similar to the hormone hCG or human choriogonadotropin. This hormone is known popularly because it is used in drugstore tests for pregnancy; if it is present, the woman is pregnant (or she has cancer). The hormone is well known in cancer research, as Livingston and Livingston explain:

> The presence of hCG in the cancerous has been observed for over 70 years. In 1902 Beard published his work on the trophoblastic thesis of cancer (Beard 1911). [The theory holds that a type of embyronic cell, the trophoblast, may remain dormant in the host organism, and when activated later in life it may produce cancer.] During the intervening years this theory has been hotly debated, as has the validity of testing the blood and urine of cancer patients for the presence of chorionic gonadotropin. Except for certain specific types of trophoblastic neoplasms such as choriocarcinoma, hybatid mole, and teratocarcinoma, the trophoblastic theory for all cancers is largely discounted. However, interest in cancer as an endocrine disease has mounted enormously in the past five years. Cancer is now being interpreted not as an invasive uncontrolled growth but as a failure of cell maturation due to the removal of growth control or to a loss or derepression of cell regulatory mechanisms. It is now postulated that the regulatory failures may be due to various types of endocrine aberrations mediated through the DNA and RNA of the cell nucleus. (Livingston and Livingston 1974: 572)

The idea that bacteria can produce a human growth hormone may seem ridiculous, but the key to understanding this claim is that bacteria produce an hCG-*like* substance. Livingston's claim that bacteria can produce an hCG-like substance was soon replicated in other laboratories, and it probably became the least controversial part of her work. In 1976 Herman Cohen and Alice Strampp of the Princeton Laboratories confirmed the bacterial synthesis of a CG-like material from the Livingston samples, and the laboratory of Hernan Acevedo in Pittsburgh, who first identified hCG in human sperm and also discovered that cancer cells in culture synthesize hCG, published confirmation in 1978.[11] Other publications found choriogonadotropinlike substances in a variety of common bacteria isolated from tissue samples from cancer patients.[12] Additional work from Acevedo's lab concluded that expression of the CG-like substance was a strain (not a species) characteristic, that not every bacterial strain from cancer patients expressed the CG-like material, and that CG-like-producing bacteria did not necessarily indicate the presence of active disease.[13]

Domingue, Acevedo, and colleagues showed that vaccines from killed *Staphylococcus haemolyticus* and *Streptococcus bovis* elicited antibodies in rabbits that were "immunologically similar to those produced in response to the whole human trophoblastic hormone," and that the bacterial CG-like material appeared to be located on the membranes of the cell wall (Domingue et al. 1986: 97). In a subsequent publication they used electron microscopy to show that there were morphological alterations in six of the nine strains of bacteria that were producing CG-like substances, and control bacteria did not show morphological alterations. One of their nonproducer control bacteria was from the urine of a pregnant woman, "containing therefore high amounts of hCG, thus suggesting that the observed alterations of the hCG-producers were not due to the trophoblastic hormone, since this control bacterium was growing naturally in the presence of a high concentration of hCG" (Acevedo et al. 1987: 790). They add:

> It is also possible that alterations at the level of the cell wall during reversion may produce changes in the antigenic characteristics of the bacteria, leading to altered immunological responses in the host. It is now known that there are immunological specificities associated with certain CWD bacteria that are not evident when intact parental organisms are utilized as immunogens. . . . Because of these characteristics, hCG-producing bacteria as CWD variants or revertants may have pathophysiological importance especially if an association exists between these bacteria and the process of malignant transformation, since most of these bacteria have been isolated from patients with clinically manifested cancer. (Acevedo et al. 1987: 790)

Acevedo's research in the 1990s has moved away from the bacterial expression of CG-like substances and back to his original interest in cancer cells and their relationship to CG (Acevedo et al. 1995a, 1995b; Krichevsky et al. 1995). Continued research from his and other laboratories has shown that the expression of membrane-associated hCG is common in cultured human malignant cells, but not detectable in nonmalignant tissue. This research has fueled continued speculation about the similarities between embyronic and cancer cells. It is suspected that placental hCG protects the embryo against rejection by the mother's lymphocytes. Likewise, hCG (particularly the beta chain) as a continuous layer on tumor cells may play a similar role of preventing immune system recognition, perhaps because hCG's negative charge repels lymphocytes.[14] In late 1995 Acevedo, Tong, and Hartsock published strong evidence that a wide variety of cancer cells express hCG-beta on their membranes; this publication was greeted by some media attention and a very favorable editorial in *Cancer* with the enthusiastic title, "Have We Found the 'Definitive Cancer Biomarker'?" (Regelson 1995). Furthermore, because hCG can stimulate tumor cell growth in vitro, it may turn out to be more than a tumor marker.

Vaccines for birth control using hCG have been widely tested, and some research suggests that similar vaccines for hCG may have some impact on tumor genesis and/or growth. In 1982 Acevedo and colleagues showed that rats inoculated with the beta subunit of choriogonadotropin developed antibodies to CG and showed fewer instances of cancer than controls, which did not show CG antibodies (Kellen, Kolin, and Acevedo 1982; Kellen et al. 1982). Vaccines using CG have begun to be tested on cancer patients with nontrophoblastic cancers; preliminary trials by Pierre Triozzi of Ohio State University and colleagues suggest that the vaccine "is well-tolerated and has biological activity in patients with cancer" (Triozzi et al. 1994: 1447). Injections of hCG into Kaposi's sarcoma lesions also appear to be effective (Gill et al. 1996).

It is possible, then, that the antiserum developed by Glover worked because it contained CG antibodies rather than antibodies to a specific cancer bacterium or bacteria, or that the bacterial vaccines of Livingston, Diller, and Seibert were successful because they stimulated CG-antibody production (Netterberg and Taylor 1981: 82). The research of the 1980s suggests that bacteria do indeed produce a CG-like substance, so much so that the term "hCG" has sometimes shifted to "CG" (therefore dropping out the "human"). Furthermore, Domingue, Acevedo, and colleagues noted that the bacterial production of CG-like material suggests that CG is a "primeval molecule" or that it arose in bacteria and vertebrates by convergent evolution. They reject as unlikely the counterargument that CG genes in bacteria emerged from genetic exchange with already transformed malignant host cells because the two subunits of hCG are made by separate genes and because "specific enzymes are needed to add the sugar moieties."[15] Subsequent analyses have led other scientists to the following conclusion: "A gene with homology to the hCG beta-subunit family exists in bacteria, suggesting that either such genes evolved independently in bacteria, or that the hCG gene family is older, in an evolutionary sense, than is currently believed."[16] Assuming that bacteria do produce a CG-like material independently of malignant host cells, then autogenous vaccines—that is, vaccines more like the Livingston type than the Glover serum—could be specific to the CG-producing bacteria and therefore work by eliminating the CG-producing bacteria rather than counteracting CG-production. However, given the wide number of strains of CG-producing bacteria, other mechanisms may be more important (such as stimulating the immune system to destroy cells with CG antigens on their cell walls).

To summarize the argument to this point, the claim that there is no single cancer organism appears correct in light of the best available current evidence. However, I would suggest that the power of the argument to defeat the bacterial-infection theory is weakened by both the possibility that standard bacterial classifications may be overly rigid and by the subsequent findings that bacteria from malignant tissue have been found to express molecules similar to human choriogonadotropin. The linkage of the bacterial-etiology theory with CG research helps overcome the inconsistency problems discussed earlier, because it translates the theory into contemporary immunology and molecular biology. At the same time, the already-modified bacterial-etiology theory has been modified again in order to recognize that there is no specific cancer organism but only specific bacterial functions (such as CG production) that could provide an etiological role for common CWD bacterial pathogens in tumor genesis and/or growth.

The Pragmatic Argument I: Rarity

I divide the pragmatic argument against the bacterial-infection theory into two parts, the rarity argument and the vaccine-inefficacy argument. The rarity argument is as follows: even if one were to accept the possibility that bacteria play some role in the human cancer, is it not likely that the role is an extremely minor one? Therefore, if the role is a minor one, might the theory be ignored as tangential and might the research program legitimately be excluded from funding? After all, according to current estimates viruses probably "cause" fewer than 10 percent of human cancers. Bacteria are likely to play a smaller role. Therefore, it would be a waste of funds to invest in a field that is likely to have a payoff only in a small percentage of human cancers.

The first counterargument could be framed as follows: if bacteria play a role in tumor genesis via production of a CG-like substance, and if CG turns out to be the definitive tumor marker that the *Cancer* editorial suggested, then it is possible that the role of bacteria in tumor genesis is substantial. This argument is elegant but it involves several unconfirmed assumptions.

A second counterargument is more empirical: it evaluates the frequency with which bacteria have been isolated from cancer tissues. For certain types of cancer, researchers seem to have isolated bacterial cultures with relative ease and frequency. For example, a series of studies in the 1960s suggested that mycoplasma could be isolated relatively frequently from leukemic bone marrow or tissue cultures inoculated with human leukemic bone marrow.[17] A more long-lived and consistent pattern is with Hodgkin's disease. Early in the century German researchers isolated bacteria from the glands of Hodgkin's patients, and in subsequent years the University of Wisconsin pathologist C. H. Bunting replicated their work and found highly pleomorphic bacteria in cultures from Hodgkin's patients.[18] Bunting also used injections from the Hodgkin's patients' cultures to produce a blood leukocyte picture in monkeys that he claimed was similar to the Hodgkin's picture in man. Other researchers continued to get similar results into the 1930s and 1940s (L'Esperance 1931; Mazet 1941). A setback for this research suggested that lymph glands may contain a normal flora, thus making it difficult to assess the pathological role of bacteria cultured from lymph glands (Adamson 1949). However, other scientists subsequently found bacteria in blood cultures from Hodgkin's patients, suggesting diseased lymph nodes, and they showed that the nitroblue tetrazolium test was positive in all cases of Hodgkin's disease, suggesting infection.[19] Researchers such as Alan Cantwell, Jr., have kept this research tradition alive by continuing to find bacteria in Hodgkin's cultures.[20]

Sarcomas may represent another type of cancer for which bacterial infections may play an important role. The range of transmissibility of the Rous sarcoma virus across species remains unresolved, but some of the Livingston network research involved murine sarcomas and therefore suggested transmissibility to mammals (e.g., Livingston and Majnarich 1986; Diller and Donnelly 1970). Livingston and colleagues believed that the microorganism could be transmitted to humans through undercooked chicken (Livingston 1984: 80; see Gross 1983: ch. 7). Furthermore, Coley's toxins are thought to be particularly effective in sarcomas (Wiemann and Starnes 1994).

Other researchers claim to have isolated bacteria from a much wider range of cancers. Most of the researchers cited here, such as Glover and Livingston, were working with a variety of malignant tissues, and their reports suggest ubiquity rather than rarity. Glover, Scott, and colleagues claim that they "have been able to isolate and to culture a morphologically similar, Gram-positive pleomorphic organism from every type of malignant growth with which we worked, including human carcinoma, mouse carcinoma, rat carcinoma, human sarcoma, rat sarcoma, and Rous chicken sarcoma No. 1" (1926: 50). Gerlach (1948) reported successful cultures from over one thousand samples of blood and tissue from humans and animals (Diller 1962a: 203). Likewise, Livingston claimed, "On examining all kinds of cancer tissue obtained directly from the surgeon in the operating room to insure sterility and absolute freshness, I found that a similar microorganism [to that of her scleroderma work] was present in all of them" (1972: 16). Seibert, who entered the field partly because of her experience with demonstrating contaminants in tuberculosis research, wrote, "We found that we were able to isolate bacteria from every piece of tumor and every acute leukemic blood specimen that we had" (1968). Diller provides another suggestion of frequency in her study of one hundred female ICR/Ha strain mice. At periodic intervals, including at death, she bled the mice and attempted to culture pleomorphic bacteria. Of the fifty-six mice that died from tumors, forty-nine had successful cultures, and of the forty-four that did not die from tumors, only seven had successful cultures. Of the seven that were apparently false positives, three were killed by cage mates (suggesting possible infection) and the remaining four died from pulmonary congestion and edema (suggesting other infections).[21]

The ubiquity argument is bolstered by the existence of an interesting trail of publications by researchers who were apparently unfamiliar with the main lines of the North American and European networks on bacterial variation and cancer, yet they have published articles that independently suggest they have found indications of bacteria or fungi in malignant tissue. For example, John Gregory—a doctor who appears to be unaware of the Glover, Rife, and Livingston research—compared filtrates from samples taken from surgery of one thousand malignant tissues with another sample of one thousand benign tissues. Using an electron microscope, he concluded that "spherical viruslike bodies .1μ in diameter were found in 100% of the malignant tissue but never in the benign tumors or normal tissue. These objects have cell detail, including cell wall, nucleus, and cytoplasm" (1952: 20).[22] In short, the available research suggests that when researchers are looking for bacteria or fungi and when they use electron microscopy or difficult—but—appropriate culturing methods, they are likely to find evidence of microbial pathogens in large numbers of malignant tissue cells. Thus, there is credible empirical evidence against the rarity argument. Certainly, one would want much more research with modern techniques in order to evaluate the issue properly. It is possible that future research will conclude that bacteria are only important etiological agents in some cancers, such as leukemias, lymphomas, sarcomas, and cancers of the digestive tract. However, bacterial vaccines and sera may have some efficacy for a wide range of cancers.

The Pragmatic Argument II: Inefficacy

The vaccine-inefficacy argument has two general types. First, if one does not accept the bacterial theory, then bacterial therapies of the Glover or Livingston type are classified as nonspecific immunotherapies similar to the BCG tuberculosis vaccine or Coley's toxins. Presumably, these therapies operate by stimulating cytokines (regulating molecules) that stimulate the immune system. In general, nonspecific bacterial immunotherapies have a checkered history of successes and failures, and as a result many researchers have questioned the value of pursuing them. Therefore, one could extend this argument to the Glover/Livingston therapies and argue that they are likely to have a similar history of spotty successes.

A counterargument might begin with the claim that nonspecific bacterial vaccines such as Coley's toxins are of questionable efficacy. Research compiled by Helen Coley Nauts and others suggests that claims of a lack of efficacy for Coley's toxins have been overstated, and there is sufficient accumulated data on outcomes to suggest that five-year survival rates for patients treated with Coley's toxins under the proper conditions will be equal or better than comparable five-year survival rates for conventional therapies.[23] A second step to the rebuttal is to recognize that the argument assumes the conclusion: that the bacterial theory is incorrect. If the theory turns out to be correct, then vaccines made from bacteria cultured from the patient could be classified as specific, and consequently the treatments could be in theory more efficacious than other nonspecific bacterial therapies. Thus, if the bacterial etiology theory is correct and if autogenous vaccines are used, the comparison with the mixed outcomes of other nonspecific bacterial immunotherapies may not be appropriate. Furthermore, as Coley noted long ago, vaccines developed for one type of bacterium, such as tuberculosis, can be effective for diseases caused by other bacterial agents, such as leprosy (Coley 1931: 614). Thus, it is possible that general bacterial vaccines such as Coley's toxins may operate to some extent by affecting cancer-associated bacteria directly or by stimulating CG-antibody production.

The second type of vaccine inefficacy argument is based on actual trials in humans and animals. Unfortunately, the quantitative data for humans are not conclusive. One source of data is Glover and White's 1940 report, which to my knowledge has never been analyzed. The report includes follow-up statistics on the fifty cases first presented in 1926. Because he used those cases to show that the serum worked, he provided no comparative evaluation of successes versus failures. Thus, the fifty-case pool shows only the frequency of long-term successes in cases that had already achieved short-term success, and therefore these statistics will not be considered here. A second set of statistics is drawn from 278 cases that "represent groups of cases as they occurred in a few of our longer established clinics" (1940: 22). He adds that "cases in these clinics that were symptom-free for only a few months were not included because of the insufficiency of the time element for proper evaluation of results" (22). Although this statement is not entirely clear, I interpret it to mean that he did not include remissions that, at the time of data collection, had only been achieved for a few months. Had he included these statistics, his success rate would have been better.

Within the pool of 278 cases, the breast-cancer cases form the largest group that can easily be compared with a contemporary diagnostic category. Glover divides the ninety cases of breast cancer into a group of thirty-two operable and fifty-eight inoperable cases. In all but five cases pathologists verified the tissues as malignant (Glover and White 1940: 38). The remaining five cancers were in patients who refused biopsy but for whom clinical examination revealed "the characteristic lesions of malignancy, so that no reasonable doubt was left in the diagnosis" (38). Glover concludes that all ninety cases were indeed correctly diagnosed as breast cancer.

In the operable cases, most of the patients had no previous treatment, but in the inoperable cases many of the patients had experienced recurrences after surgery and/or radiotherapy (see Table 1). Those with previous treatment had suffered a recurrence on the average of nine to fourteen months after treatment. Although the doctors used surgery together with the serum in most of the operable cases, Glover was an early advocate of

TABLE 1.
Glover Serum Outcomes for Breast Cancer

Operable (32)	Inoperable (58)
Previous treatment:	Previous treatment:
30 none	19 none
2 surgery	21 surgery
	15 surgery and radiotherapy
	3 radiotherapy
Current treatment:	Current treatment:
26 serum and surgery	58 serum only
6 serum only	
Outcome:	Outcome:
8: 10–12 years	8: 10–12 years
12: 5–9 years	5: 6–9 years
9: 1–4 years	12: discontinued
2: regression and recurrence	1: regression and recurrence
1: cerebral embolism	4: infections
	2: diabetes
	1: fractured neck
	22: death
	3: no information
Rough Five-Year Survival:	Rough Five-Year Survival:
24.5/(32–1) = 79%	13/(58–12-4–2-1) = 33%

SOURCE: Glover and White (1940: 38–43).

conservative surgery: "Performance of the more radical operations with manipulation of the malignant areas has a proneness to disseminate the disease, and . . . our results have been best when serum alone was used or when we have used conservative surgery preceded and followed by serum therapy" (Glover and White 1940: 39).

To interpret the outcomes properly, the figure of, for example, eight patients alive at ten to twelve years does not mean they only lived ten to twelve years. Rather, they were still alive and healthy at the time when data was collected. In other words, some received treatment only four years or one year prior to the collection of data. Of the remaining patients, two had complete regression, refused additional serum treatment even though the blood culture was positive, and then later had metastatic recurrences. One patient died from a suspected cerebral embolism while undergoing a simple surgical excision. Of the six patients treated only with serum, two each were among the three surviving groups.

The results for the inoperable cases were not as promising. Twenty-two patients, all of whom were at advanced stages of the disease, died. Glover comments that for them "relief from pain without the aid of narcotics was noted in the majority of the cases" (Glover and White 1940: 43). Of the other patients who were not in the survivor group or the group that discontinued treatment, four died from infections before the treatment was completed, two died from diabetes but with evidence of some tumor regression, one died from a fracture of the neck (but with no evidence of metastases at the site of the fracture), one died after recurrences from complete regression (she refused additional serum after regression even though blood cultures remained positive), and no information is given for the three remaining cases.

Some of the case histories provide remarkable evidence for long-term, complete regression of tumors among very advanced patients. One forty-four-year-old patient began with symptoms of pain and swelling in the left breast, and she was treated in 1918 by an excision of the lump from her breast. When the pain returned and more nodules were detected, the left breast was removed in 1920, and pathologists reported adenocarcinoma of the breast. A new tumor in the right breast resulted in removal of the right breast in 1921, with the same pathological report. Shortly after this operation the patient experienced shortness of breath, chest pains that radiated into the arm and neck, weakness, and loss of weight. By the end of 1922 she was bedridden and given narcotics to alleviate her pain. Additional tumors were detected on the ribs and left arm, and the lower border of the liver was hard and nodular. Dr. Duffy of Troy, New York, administered the Glover serum, and within six months she was ambulatory enough to travel 150 miles by train for additional treatment. Glover continues:

> At the end of one year of treatment no evidence of any carcinomatous involvement could be found. The patient had gained thirty-three pounds in weight and was leading a normal existence. The blood culture was negative.
>
> In September 1925, the duties of teaching a large fourth grade class were resumed, and, except for seasonal vacations, have been performed without any break up to the present time. At this time of writing the patient is approaching her sixtieth year without recurrence. (Glover and White 1940: 44–45)

Of course, case studies such as this fall into the category of anecdotal research. The advantage of Glover's research is that it provides some basis for a quantitative assessment of outcomes. Returning to the statistics, one might ask, how does Glover's serum compare with today's statistics? Five-year survival rates for breast cancer among white women in the United States have changed from about 63 percent in 1960–63 to 82 percent in

1983–90.[24] The gains are probably optimistic and due mainly to earlier diagnosis than to more effective treatment. To compare with Glover, let's take his more optimistic statistics on operable patients, because the operable cases included twenty-one patients who had local metastases and whom some of the clinicians thought should be classified as inoperable. This pool is therefore closer to the current patient population upon which the national average figures are based. Assuming that at least half of the nine patients in the one-to-four-year group make it to five years, then the protocol has a five-year survival rate of twenty-four-and-a-half patients of a group of thirty-one (dropping the patient who died of the embolism). This gives a rough five-year-survival figure of 79 percent, which is comparable to today's optimistic figures. Glover also appears to have had a ten-year survival rate of over 60 percent, that is, a success rate that is closer to what many think is a better estimate of the five-year survival rate for the national average today.

John White (1953), a doctor who used the Glover serum at his clinic in Malone, New York, published a second report on another one hundred cases. For breast cancer, the largest group, he had a five-year survival rate of 65 percent and a ten-year survival rate of 57 percent. The statistics are very rough because he apparently lost track of some patients prior to the five- and ten-year markers, and some of the deaths may not have been cancer-related. The statistics on the Glover serum can only be rough estimates by today's standards of measurement, but they are all that is available so they are of some value if one understands their limitations. Given the likelihood that diagnoses in the 1920s and 1930s occurred at much later stages of cancer than today, the statistics suggest that the Glover serum may have been efficacious even when compared with today's success rates. Furthermore, his serum treatment held out the possibility of significantly lower disfigurement and higher quality of life to the extent that it could reduce or replace other treatment modalities. Only additional testing would have answered this question.

Regarding the Livingston vaccine, there is now an *apparently* better measure of efficacy in human populations than the very rough measures that I have constructed for the Glover serum. There is finally a matched cohort study that has been designed in accordance with the methodological standards of late twentieth-century medical research. Barrie Cassileth and colleagues (1991) compared matched pairs of patients under a mixed conventional/Livingston protocol at Livingston's clinic in California and under conventional treatment (including interleukin-2) at the University of Pennsylvania. The study showed no difference in mean survival time, and all but one patient (a Livingston patient) died within three-and-a-half years after diagnosis. Furthermore, quality-of-life measures assigned higher scores to the conventional patients at the beginning of the study, and the gap between the two therapeutic protocols remained consistent throughout the study. This study could therefore be interpreted to provide strong evidence against the vaccine-efficacy argument for the Livingston vaccine.

However, there are some additional considerations that need to be taken into account before full evaluation of the study is made. It is not irrelevant to include in the interpretation of the findings of the study a statement in the pro-alternative publication *Cancer Chronicles* that Cassileth is a member of the American Cancer Society's Subcommittee on Questionable Methods of Cancer Management, a continuation of the older Subcommittee on Quackery. Because of that membership Cassileth was part of a controversy among members of the advisory board of the Office of Alternative Medicine, whose cancer-activist members wanted her to resign from one or the other.[25] To her credit, she has advocated controlled studies of alternative therapies, a position which itself is controversial among the cancer establishment. On the other side, Cassileth was quoted in a *U.S. News and World Report* article as criticizing some OAM advisory board members for using their affiliation to promote specific alternative therapies.[26] In other words, both sides are claiming that the other has vested interests. This background knowledge leads me to read the study carefully to consider if any aspects of its interpretation, measurement, or design might have favored—perhaps unintentionally—the conventional therapies over the Livingston therapy.[27]

One possible example involves the interpretation of the quality-of-life measures.[28] The abstract of the study provides only the following summary sentence: "Quality-of-life scores were consistently better among conventionally treated patients from enrollment on" (1991: 1180). Yet, the claim that quality of life was better among the conventional patients is ambiguous. Fifty-one of the conventional patients were fully ambulatory at the beginning of the study in contrast to forty-six of the Livingston patients, and three of the conventional patients were bedridden in contrast to seven of the Livingston patients. In other words, there is a difference in the patient groups that favors the conventional protocol. The authors state that there were no interaction effects, "meaning that the quality of life deteriorated at an equal rate in the two patient groups," and a graph shows a constant and equivalent gap in quality-of-life scores over time (1991: 1183). Thus, an alternative interpretation of the quality-of-life measures is that there was no significant difference between the two because the gap remained constant over time. Furthermore, the differences between the two groups of patients as reflected in quality-of-life scores may affect interpretation of survival differences. As Michael Lerner noted: "If a reanalysis of the data in this study were to find that the difference in quality of life reflected the fact that the Livingston patients were in fact significantly sicker from the start—purely a hypothetical possibility—then the fact that they lived as long as the University of Pennsylvania patients would suggest that the Livingston regime was slightly more efficacious than conventional therapy for patients with these diagnoses and stages of disease" (1994: 330).

Finally, even if a reanalysis were to show that there is no significant difference between the therapies on mean survival time and quality of life (as measured by a constant gap), then one would probably select therapies based on cost. A vaccine/dietary treatment is much less costly than radiation, chemotherapy, and interleukin-2. The alternative suggested by the authors, that for advanced patients only palliative treatment may be the best option, is also a possibility to entertain. However, patients are likely to pursue some sort of treatment that gives them hope, and palliative treatment does not provide hope. Therefore, the study can be interpreted as favorable to the Livingston therapy because quality of life and efficacy are equivalent, but cost is lower for the alternative therapy.

Additional research would be needed to investigate some other questions that emerge from a critical examination of the design of the study. First, as Cassileth and colleagues recognize, a more convincing comparison would involve a randomized design. A better design would also measure only patients who used only the most conventional therapies (chemotherapy, radiation, and surgery) against patients who were pursuing only the Livingston protocol (rather than a mixture of conventional and Livingston protocols). I hypothesize that patients matched on quality-of-life scores prior to treatment would diverge in favor of the Livingston protocol at least on quality-of-life measures and perhaps on survival as well. It is hard to imagine quality of life being better under chemotherapy and radiation in comparison with carrot juice, organic foods, and vaccines, unless the latter protocol is so bad that patients advance more quickly to late stages of cancer.

Another design improvement would measure patients who were not at such an advanced stage of cancer, preferably patients who have received no previous treatment. Many of the alternative cancer protocols, like the conventional protocols, claim that they lose effectiveness for patients who are in the terminal phases of their illness. Regarding Coley's toxins, Guo

Zheren and Helen Coley Nauts (1991) write that the results were not as good with the bacterial vaccine when the tumors were enormous or recurrent, or after surgery and chemotherapy had been given first. In general, claims for long-term survival for both alternative and conventional treatments are built around early-to-middle-stage cases, even though case studies of terminal patient recovery can occasionally be produced for both conventional and alternative therapies. Significant differences (either for one side or the other) are more likely to emerge in designs that begin with patients who are not at advanced stages. For example, the Gerson melanoma study found statistically significant results among Stage III and IVa patients treated under their protocol in comparison with national averages, and a positive data trend for earlier stages, but the advanced patients (Stage IVb) all died (Hildenbrand et al. 1995).

Like the matched-pairs design of the Cassileth study, the Gerson study's method of comparing outcomes from its protocol with those of other studies is not as convincing methodologically as the gold standard of randomized, prospective, clinical trials. (The double-blind requirement would be impossible to incorporate if one is comparing vaccines and diet versus chemotherapy and radiation.) However, gold-standard studies are aptly named because they are very expensive. Randomized controlled trials for alternative cancer therapies would have to receive government funding because most of the advocates of alternative cancer therapies do not have the resources to break through this methodological glass ceiling. In turn funding would come only from public pressure on members of Congress, and even in this circumstance the history of the laetrile, Vitamin C, antineoplastons, and other randomized clinical trials suggests that the randomized, clinical trials of unconventional cancer therapies have been designed in ways to introduce biases against the therapies (Moss 1989). Given the severe budgetary limitations of advocates of alternative therapies, the Gerson study is a model of what can be achieved with limited resources.

The argument that outcomes studies are methodologically weak should therefore be taken with a grain of salt, because critics usually are not willing to work toward getting the funds for gold-standard studies for alternative therapies. It is equivalent to criticizing someone for not driving a Mercedes when they are on the budget of a subcompact. For that reason one should not be too critical of Livingston's qualitative outcomes analysis, which was based on one hundred charts drawn randomly from the clinic's files. The cases suggest that her immunotherapy program may have been successful for some patients (Livingston 1984).

Another low-budget alternative to randomized, clinical trials is experimental animal studies. One example is the Diller mice studies that used killed vaccines of the acid-fast or partially acid-fast stage of organisms harvested from the mouse Sarcoma 180. Using mice with a known tumor incidence, Diller found that half the control mice died within four weeks, whereas none of the vaccinated mice died within that period. The rate of tumor regression was over three times greater in the vaccinated mice than in the control mice. For those mice in the vaccinated group that eventually did not overcome the tumors, the survival time was increased (Diller and Donnelly 1970: 671). Seibert also found that a heat-killed vaccine from a filterable form of *Staphylococcus epidermidis* that was isolated from a spontaneous mammary tumor of a C3H mouse led to a statistically significant decrease in the incidence of spontaneous tumors of vaccinated versus control female mice. She believed that a more prolonged effect might be obtained with repeated booster doses, which are part of the standard Livingston therapy.[29] Livingston's bacterial antigens are also used commercially in a licensed, killed vaccine for Marek's disease (neurolymphomatosis), a neoplastic disease of chickens caused by a herpes virus.

The animal studies, combined with the design problems in the one matched-pair clinical study and some very tentative signs of success for the Glover serum, suggest that the serum/vaccine inefficacy argument remains an open question. The research mounted by Nauts as well as subsequent clinical trials for Coley's toxins also provide supporting evidence in favor of bacterial immunotherapies in general, although of course one must take into account the fact that Coley's toxins are different from autogenous bacterial vaccines. More studies would be needed to evaluate the efficacy of the Glover serum, Livingston vaccine, and related therapies based on the bacterial-etiology theory. However, these studies seem warranted given the successes of the animal studies with heat-killed vaccines and the claims that Glover and White made for the success of the Glover serum. It seems very possible and tragic that highly efficacious bacterial immunotherapies for cancer have been overlooked for decades for political reasons.

Note that advocates of the secondary-infection theory could argue that any results that suggest efficacy for a bacterial vaccine could be explained under the "cascade of cytokines" theory advocated by the Coley's toxins researchers (e.g., Wiemann and Starnes 1994). This is a formidable argument, but it could be tested by comparing nonautogenous (Coley's toxins) and autogenous vaccines (Livingston) for efficacy, production of the cascade of cytokines, and production of CG antibodies. It is possible that both mechanisms (e.g., CG antibodies and cytokines such as tumor necrosis factor) may be operant in both types of vaccine therapies. However, it is also possible that different mechanisms may be involved and that both theories (such as CG mechanism and the cascade of cytokines) may be correct to some extent. Likewise, if one were to accept the claimed efficacy of the Glover serum, any number of explanations is possible. The serum might provide antibodies to oncobacteria, but it might also work through some other mechanism of immune-system stimulation. The point is that this rather obvious and relatively inexpensive research simply is not being done, notwithstanding the billions of dollars dedicated to cancer research each year.

The Feminist Criteria

Many readers, particularly male scientists, will probably cringe at the thought of introducing feminist criteria into the evaluation of a scientific theory. However, feminist science studies analysts have developed a substantial literature that shows how scientific theories—mostly in the biological, medical, and social sciences—have sometimes been laden with gender and race biases. Thus, these considerations need to be incorporated into a fair evaluation process.

The strongest case for gender and race bias in natural scientific theories (rather than institutions) is probably in the medical and biological sciences rather than the physical or mathematical sciences. In medicine, theories related to the biology of women and people of color have historically legitimated a number of sexist and racist practices. Probably the most notorious are the nineteenth-century sciences that legitimated the exclusion of women and people of color from higher education because of the constitution of their nervous system or size of the brains (Harding 1993). There is also a long history of treating psychological disorders by removing parts of the woman's body and of medicalizing natural occurrences such as menopause and menstruation (E. Martin 1987). In cancer research and treatment, probably the most obvious area where sexist biases have been challenged is the practice of radical mastectomy for breast cancer, which only in recent years has been challenged in the United States. Because vaccines or sera have very different implications for women with breast cancer in contrast with surgery, the bacterial theory could be interpreted as more favorable to women than the secondary-infection theory.

A second type of bias is the more subtle use of implicit gendered or racial cultural categories in scientific theories. This type of bias is both more controversial and more interesting because it is applied to basic scientific theories that on first analysis may appear to be pure "representation" of the material world and free from "cultural contamination." The most relevant example for cancer research is the analysis of debates over the assignment of activity and passivity to spaces within the cell. Historically, the sperm was seen as an active agent that fertilized a passive egg, and the gendered activity/passivity relations were then mapped onto the relationship of nucleus to cytoplasm in the cell.[30] The central dogma of molecular biology—the DNA defines the structure of RNA, which governs the making of protein—continued the assignment of activity/passivity relationships by retaining the "master molecule" in the nucleus as the locus of activity in contrast with a passive cytoplasm. Some biologists, among them Ernest Everett Just and Barbara McClintock, challenged various versions of a rigid chain of command that flows from the nucleus to the cytoplasm and cell membrane.[31] They and others were pioneers in arguing for what Helen Longino (1994) calls "complexity of relationship" in their natural objects, in this case a more complex model of the cell that brings attention to the flow of information and messages inward to the nucleus.

Under older versions of the mutagenic or hereditary theory, cancer research continued this assignment of activity/passivity relations by focusing the tumor genesis process on damage to or changes in the chromosomes and later the DNA. However, gene expression has come to be seen as a more complex process, and the simple mutagenic theory has changed to allow for the expression or derepression of nuclear genes to be responses to extranuclear information. This tendency is increasingly evident as the molecular-cancer research program moves toward analyzing the failure of DNA-repair mechanisms as a major cause of cancer, rather than the simple impact of carcinogens and free radicals on nucleotides. For example, biochemist Bruce Ames (1995) has argued that foods contain a large number of natural pesticides, and therefore the cause of cancer rests in the delicate balance of DNA damage and DNA-repair mechanisms, not simply the impact of external carcinogens. In general, the molecular theory today assumes some impact on DNA expression patterns from the flow of information back from the cytoplasm and cellular environment. In other words, under the molecular theory today assignments of activity and passivity are more complicated than under a simple mutagenic theory. Nevertheless, the focus of attention remains on the DNA and its expression patterns. Sometimes one reads that the DNA "expresses" itself, whereas the environment sends "signals," like stereotypes of airspace-grabbing men and subtly signalling women. This linguistic opposition suggests that the new molecular language may continue some of the old patterns of the cultural geography of the cell.

Because the bacterial theory locates activity in a cellular pathogen, it would further complicate the understanding of molecular information pathways that lead to the transformation of a normal cell into a cancer cell. Bacteria and viruses may be located in the nucleus or in the cytoplasm or even outside the cell, but wherever they are located they complicate the older "master molecule" image of DNA-centered carcinogenesis. The proposed CG mechanism also emphasizes activity at the cell wall rather than damage to the cell nucleus.

Another candidate for implicit bias is the belief that cell-wall deficient bacteria are unlikely pathogens. I have already investigated the gendered nature of the cultural meanings associated with "cell-wall deficient/divergent" bacteria. One could then argue that interpreting these bacteria as secondary infections continues in cancer research the marginal position that they occupy in microbiology, where their pathological (active) role has been under-recognized in comparison to the recognition given to conventional bacteria with cell walls. Yet another candidate for sexist bias is the popular theory of multistep carcinogenesis that involves the model of oncogenes (like an accelerator) and tumor-suppressor genes (like a brake), which could be contrasted with the theory that it involves hCG (like an embryo or fetus). Although the two theories are not necessarily contradictory, the metaphors that are used to understand the phenomena (automobile/embryo) have gendered connotations in our culture.

Looking for hidden sexism or racism in the allocation of gendered meanings can provide a way of seeing potential biases in theories that at first appear to be transparent representations of a physical world. The counterargument is that this type of analysis confuses the metaphor with the represented object. However, advocates of the counterargument generally assume that it is possible to weed out the biases or metaphors, and consequently to produce a completely transparent representation of the world. I prefer the more sanguine view that it is probably impossible to construct a theory that has no hint of sexism or racism in it, particularly if one interprets sexism or racism loosely as the use of cultural categories such as activity and passivity. Although some theories may be less racist or sexist than others, and it is worth including these grounds as one set of criteria for theory evaluation, it is doubtful that a utopian state will ever be reached in which all forms of buried or cryptic bias will be eliminated from scientific theorizing. Furthermore, the utopia of a completely culture-free science may not even be desirable, for scientific theories depend on the culture-imbued imaginations of scientists and their culture-laden metaphors. In short, culture in science both introduces biases and produces possibilities of new ideas. The point is to examine theories for their potential use of gender- or race-associated cultural categories, and to use this form of cultural critique to pose alternatives that in turn need to be cross-checked against the other criteria of accuracy, consistency, and pragmatic value.

Antisexist/antiracist criteria could play an important role if they revealed very obvious forms of bias, such as the attribution of intelligence differences based on presumed brain-size differences or the removal of women's body parts based on dubious psychobiological theories. In the case of the bacterial-etiology theory, I see no evidence for this degree of obvious bias. Probably more important than the activity-passivity argument is the possibility of using vaccines as an alternative to surgery, as in some of Glover's breast cancer patients.

Regarding the relationship among the four groups of criteria, most philosophers—including the feminist philosopher Longino—agree that some version of accuracy is the most important. Consistency criteria probably are most useful for preliminary filtering of highly inconsistent and unlikely theories, and for evaluating theories that approximate evidential indistinguishability. Pragmatic criteria will always be important given the embeddedness of science in a society interested in research applications and facing the economic problem of scarcity of resources. Finally, given the history of racism and sexism in biological theories, the new criteria introduced by Longino (and drawn from other feminist theoreticians) add a valuable perspective to the evaluation of a theory. I have reoriented the feminist criteria toward an examination of racist or sexist bias (and of course the list could be expanded to include other "isms"). These criteria serve as a useful part of the overall evaluation of a new theory that will allow a check on possible hidden biases that can easily be ignored or forgotten.

Conclusion

The original theory that a single, pleomorphic microorganism is the cause of cancer is no longer viable. The Glover-Rife interpretation of chemical and radiation carcinogenesis as precursors to microbial infection may turn out to be acceptable in some cases, but there has been too much subsequent accumulated evidence to suggest that a single microorganism is the unique cause of all mammalian cancers. The single microorganism theory that most members of the research tradition supported is no longer tenable in light of the research of the 1980s that used DNA-DNA hybridization techniques to identify the bacteria as known species categories. However, as Domingue noted, current classification categories in microbiology may be too rigid and unable to encompass the possibly wide-ranging genetic exchanges that may be going on in pleomorphic bacteria.

Although the original theory is no longer tenable, there may be some life left in a modified version of it. Bacteria can change forms in response to new cultures and host conditions, just as theories can be modified in light of new data. The modified theory holds that bacterial infection is not the only proximate cause of cancer (as opposed to distant causes such as poverty) and that there is no single pleomorphic oncobacterium responsible for cancer genesis and growth. The modified theory recognizes that while there are other proximate causes of cancer (e.g., chromosome damage from chemical carcinogens, inherited genetic susceptibility), bacteria could still play a nontrivial role in tumor genesis and/or growth, perhaps only in some types of cancer.

The modified theory can explain the patterns in the CG studies of the 1980s discussed above, as well as the fulfillment of Koch's postulates and the apparent success of at least some of the sera and vaccines. The interpretation that CG-like-producing bacteria do not acquire the function from genetic exchange with already-transformed host cells is crucial to maintaining the theory in its current form. However, as any good student of science knows, the bacterial-etiology theory would not fall apart if large amounts of evidence were mounted against the proposed CG mechanism. Because there may be other mechanisms by which bacterial pathogens could contribute to cancer, the theory would only have to be changed again. One example is hormonal deregulation through the bacterial production of a substance similar to actinomycin (Livingston and Alexander-Jackson 1965a: 859; Livingston et al. 1970). Nevertheless, unless strong evidence were mounted in favor of an alternative mechanism, the failure of the subtheory on the CG mechanism would make the bacterial-infection theory much less appealing. In other words, there is still a degree of overall falsifiability involved. When a theory obtains a high anomaly load and a large number of ad hoc subtheories to cover the anomalies, people begin to abandon ship. However, the modified theory still appears to float. Indeed, it has led to some successful, surprising predictions.

As part of a broader research program, the theory also appears to have value on pragmatic grounds. If research under the theory were funded, the development of new vaccines and/or sera may prove successful for human cancer patients. The risk of failure is offset by the lower cost and higher quality of life expected from a vaccine treatment in comparison with surgery, radiation, chemotherapy, and some of the specific immunotherapies. The theory may also aid efforts at early detection and prevention by providing a framework for culturing CG-producing bacteria from blood or testing directly for antibodies. As a mechanism for early detection, screening for CG antibodies might prove superior in cost, comfort, and risk to current protocols such as mammograms or prostate exams. Finally, the theory may provide some new ideas for research both on CG and on antibiotics as a chemotherapy for cancer.

The bacterial-etiology theory, then, is not kooky, wild, or quackery. Rather, over the decades it has guided many researchers who have amassed a credible body of evidence to support it. The theory may ultimately be rejected in favor of the secondary-infection theory, but I would suggest that now is probably the worst time to do so, because the theory may be of some use in understanding the apparent advances that are being made in the CG research. In light of the evidence on the production of substances similar to CG by bacteria associated with tumors, the secondary-infection theory could be maintained by arguing that CG production occurs after host-cell genes are switched on in a process independent from microbial pathogens. The evidence that bacteria cultured from cancer tissues produce a CG-like substance, and that bacteria independently possess the genes to express the material, could be answered by arguing that tumor initiation begins in the host-cell genome but it subsequently triggers bacterial expression of CG-like substances that contribute to tumor promotion. At this point, however, the lines between the bacterial-etiology theory and the secondary-infection theory have become blurred. It is possible that the host-cell genome and bacterial pathogens are *both* expressing hCG and CG-like substances. In other words, the controversy could be settled by accepting a new, hybrid theory. I suspect that this may happen someday, but making long-term predictions for science and society is like making long-term predictions for the weather.

5

Policy Cures: Forging a New Cancer Agenda

Whatever the status of cancer as a disease, it also represents a pressing political problem. In the United States the National Cancer Institute alone spends about two billion dollars per year on cancer research, and the overall annual cost of cancer to the U.S. economy exceeds $100 billion (Brown 1990). Cancer has overtaken heart disease as the leading cause of death in women aged thirty-five to sixty-four, and it is the second leading cause of death in all other categories except men aged fifteen to thirty-four (Davis, Dinse, and Hoel 1994a; Wing, Tong, and Bolden 1995: 18–19). Although some wish to assign the bulk of the risk factors to smoking and diet, there is growing evidence that environmental carcinogens play an important role.[1] If this evidence is correct, then efforts at prevention will need to be considerably more complicated than antismoking and dietary campaigns. Some reformers have therefore called for a restructuring of the NCI budget to focus more on prevention and environmental carcinogens (Epstein et al. 1992).

Prevention is certainly the key to ending the rising tide of cancer. However, for the millions of people who have cancer or know someone who does, prevention is not a viable option. When their doctors do not give them good chances for survival with conventional therapies, or when the side effects of conventional therapies begin to look worse than the disease, they begin to explore alternatives. Discussions of the reform of the National Cancer Institute budget, and that of affiliated private organizations, therefore need to focus not only on prevention and environmental carcinogens, but also on the evaluation of alternative therapies.

Notwithstanding the lack of official support for alternative cancer therapies, the American people are examining and trying alternative cancer treatments on their own. The decision to seek such treatments is part of a larger pattern of recourse to medicine not sanctioned by the medical profession and insurance providers. According to a national survey published in 1993 in the *New England Journal of Medicine,* more than 34 percent of the American people had resorted to at least one unconventional therapy in the previous year.[2] Estimates for the use of alternative therapies for cancer in the United States run from 10 to 50 percent of the six to ten million Americans who are living with cancer at any given time; even the conservative estimate of 10 percent suggests the vast size of the phenomenon (McGinnis 1991). Because users of alternative medicine tend to be better educated than the population as a whole, it is likely that as the population becomes more educated about treatment options, cancer patients will tend to use more alternative therapies.[3] It is also likely that the use of alternative therapies will increase as state and federal officials relax regulatory barriers that have made some alternative therapies either difficult to find or illegal.

Elected public officials know how to count votes. Many are aware that a quiet revolution has been going on in the public's approach to medicine. The change is not a temporary one; it reflects a number of structural transformations in the society and economy. Demographically, the population as a whole is aging, but there is also a cohort effect as the baby boomers become senior citizens, refuse to go gently into the night, and promise to leave a legacy in the "medical Vietnam" of the war on cancer that echoes their legacy in the first Vietnam.[4] Another factor behind the quiet health-consciousness revolution is that there is more information, and more readily available information, than ever before. Health-food stores, the vitamin-and-supplements industry, direct-mail promotions, small presses, patient-support groups, and mass-circulation health magazines all provide ready sources of information that was much less accessible thirty years earlier. Web sites, discussion groups, and other electronic resources on alternative medicine emerge every month.

There is also a profound change in the relationship between patients and doctors that probably has its roots in the women's health movement and the AIDS movement (Treichler 1991). However, that change has now become part of the general culture, particularly among college-educated patients. Increasingly patients work with their doctors on their disease as a team, rather than as servants who passively carry out the doctor's orders. Finally, as immigrant populations have become increasingly diversified and as ethnic groups have become more self-conscious, some alternative therapies have become popular as expressions of ethnic identity. In short, the profound social and demographic changes that underlie the new pluralism of the medical system are structural and long-term. Alternative medicines, however one defines them, are not a fad that will go away. Rather, they represent a diversification of the medical system in which new ideas will be sorted out, sifted through, and officially sanctioned to varying degrees.

The changes in public consciousness are increasingly being translated into public policy, often with bipartisan support. Many important changes have already happened, most of them in the 1990s, but many other reforms are necessary. The two major areas of policy reform involve rights of access and rights to information.

Rights of Access

Probably the most important area of policy reform is a change that was advocated as long ago as the writing of the Constitution, when Benjamin Rush is alleged to have attempted to add freedom of medical choice to the First Amendment.[5] Two centuries later his concern about medical freedom has proven prophetic. Several states have approved legislation that allows all qualified health-care practitioners to use so-called unconventional therapies, provided that the condition is life threatening, the treatment or procedure is not dangerous to the patient, and the practitioner discusses all options with the patient. At the federal level, the "Access to Medical Treatments Act" (HR-2019, S-2140) would provide doctors with similar protections should they decide that some unconventional or controversial therapies are efficacious and safe (Sale 1995).

A second and much more controversial area where reform is needed is the Food and Drug Agency of the U.S. government. The policy dilemma revolves around the tradeoff between two goods: the public's right to medical choice and freedom, and the state's interest in protecting its citizens from charlatanism, quackery, and unsafe products. In the United States, a country known for its emphasis on individual rights and freedoms, the tradeoff has for many years been tilted in favor of the state's interest, whereas in other countries, such as Germany, there has been greater tolerance of at least some alternative medical practices. As documented in the monthly issues of the *Townsend Letter for Doctors and Patients,* the FDA's policing power has often been directed against alternative health-care practitioners and the nutritional-supplements industry, resulting in raids that seem inconsistent with general democratic values. Legislation of the type proposed in the previous paragraph represents one modest step toward preventing the FDA from abusing its compliance powers. However, the legislation is directed more toward doctors and health-care professionals, and it would not prevent FDA abuses directed at the nutritional-supplements industry. Therefore, some health-care reformers have suggested putting curbs on the FDA's policing power.[6]

Clearly, it will prove desirable to retain some policing functions for the FDA in cases of an immediate public danger posed by contaminated food, drugs, and supplements. However, legislative reform is badly needed to restrict the agency's enforcement power to prevent unwarranted raids of the clinics of holistic doctors, their patients, and supplement companies. In some cases, supplements manufacturers have been virtually shut down, have had records and products confiscated, and have been unable to find out who accused them of having contaminated products. Instead of allowing the FDA this kind of Gestapo-like power, federal law needs to restrict the FDA to recalling unsafe products, rather than banning entire categories of products and closing down entire companies. In turn, supplements manufacturers need to have the right to know who has accused them, and to be able to contest claims that recalled products are indeed unsafe. Furthermore, if court costs from raids of holistic doctors, patients, and supplement companies exceed a specified amount, there should be an automatic trigger that would provide for government support of the defendants' legal costs. Much of the repression of alternative medicine operates through a war of legal costs, in which defendants cannot match the deep pockets of government prosecutors associated with the FDA or with state-level medical associations.

Another major regulatory problem area involves the classification of vitamins, minerals, herbs, amino acids, enzymes, and other food supplements. The FDA has consistently tried to regulate food supplements as drugs or chemical food additives rather than as foods. If successful, this shift in classification would require that companies engage in costly studies to prove that their supplements are safe and effective. In effect, the transformation would destroy much of the supplements industry, probably to the great benefit of the pharmaceutical industry. In 1994 Congress passed the Dietary Supplement and Health and Education Act, but this act is problematic for a number of reasons, among them the barriers between supplement companies and health claims (Sale 1995). The result is the rather bizarre situation in which health-food stores must keep the literature on supplements separate from the supplements, and consumers have to sift through complex literature in order to find out information about particular supplements.

The basic components of a good law are the following: it would allow food and supplement companies to make truthful, nonmisleading health claims about their products; it would prevent the FDA from classifying natural food products and dietary supplements as drugs; and it would clarify the safety standard for dietary supplements to prevent unreasonable interference from the FDA with the public's right of access to supplements. The nonmisleading claims clause should be worded to allow supplement manufacturers to provide brochures with their products that discuss scientific research that they believe supports their health claims. This could include, for example, discussions of the peer-reviewed literature on dietary supplements and herbs for use in cancer prevention and therapy. Likewise, the safety standard needs to be clarified so that products are not permanently banned based on one or two bad batches, as in the case of L-tryptophan, which some claim was banned because it provided a low-cost, natural alternative to antidepressants.

Additional legislation may be beneficial to protect consumers of vitamins and food supplements. I suggest legislation that would not merely allow but require supplements companies to undergo periodic audits of their supplements from outside, private companies (but not the FDA) in order to ensure that the vitamins, minerals, or other nutritional agents are present in the supplement in the quantity claimed. The better companies already undergo such audits and make them available to the public; this should become industry standard. Another audit should demonstrate that supplement pills are actually digested rather than passed out in the feces.

In addition to changes that allow the supplements industry to make truthful health claims for supplements, changes in the drug-and-device approval rules are necessary. After 1962 drug manufacturers were required to prove both safety and efficacy. Because the standards for establishing efficacy are so onerous, the current cost of getting drug approval through the FDA is estimated at one hundred to five hundred million dollars, and it takes over ten years. The result is that new drugs and devices need to have the support of companies that expect to recuperate their costs through substantial, long-term future profits. Even for the major pharmaceutical companies, the approval process seems Byzantine, and they have given substantial funds to conservative think tanks that have produced studies in favor of simplifying the process and making it less costly (Drinkard 1996).

The situation is even more difficult for many of the alternative cancer therapies that rely on vitamins, herbs, bacterial vaccines, or other natural and/or nonpatentable products. Products that cannot be patented are poor candidates for private research, because the funds necessary to invest for approval of their use as drugs cannot be recuperated later in future profits. The FDA approval process has become a classic case of a barrier to entry in an oligopolistic market. The research that would allow natural or nonpatentable products to pass through the FDA drug-approval process is not getting done because there is no profit incentive. If, for example, one company were to invest several hundred million dollars to get a natural substance approved as a cancer drug, all other companies would benefit as free riders.

Several suggestions have been made to remedy this glaring problem in the drug-and-device approval process. Whereas some reformers have advocated abolishing the efficacy requirement completely, others suggest reducing it dramatically. In testimony before the Subcommittee on Government Oversight of the House Commerce Committee, attorney Richard Jaffe suggested that "the level of amount of proof needed to establish the efficacy of the drug or device should be directly related to the product's safety" (Jaffe 1995: 16). In other words, products with low demonstrated side effects and high safety should not have to pass the same efficacy requirements as drugs and devices that have high risks and side effects. Additional criteria for lowering the barrier to entry could include much lower efficacy standards if the substance is being used for chronic diseases and if there are no proven cures for the disease (as in the case of cancer).[7] A reasonable low efficacy standard in this situation would be a safe product that demonstrates apparent efficacy in related diseases or animal experiments. By this standard, a competent physician could introduce a safe therapy into a treatment program for a patient with cancer or another chronic disease. One step in the right direction is that in 1996 the FDA shortened approval periods for cancer drugs. However, although the changes may be helpful for the pharmaceutical industry, they do not meet the problem of regulatory barriers against nonpatentable products.

Medical writer Robert Houston has proposed additional policy changes for the regulatory process. He has argued that the FDA's record on unconventional cancer therapies reveals so much negativism and obstruction that it may be necessary to establish "a separate agency, with authority to overrule the FDA, to assess the safety and efficacy potential of alternative therapies, orphan drugs, and unpatentable natural agents outside of the expensive IND/NDA procedures of the FDA" (1989: 51). Although this proposal may not be very welcome in a budgetary climate that has focused on eliminating agencies, it is feasible and relatively inexpensive to have some sort of external appeal process for FDA rulings, such as a jury of relatively neutral peers (with selection procedures similar to those of other juries that allow both sides to exclude apparently biased jurors). Furthermore, Houston's second proposal would also be relatively inexpensive: "automatic demotion and possible criminal penalties may be imposed on government personnel who knowingly or prejudicially misrepresent or mistreat alternative therapies or their proponents" (1989: 51).

Finally, access issues are severely restricted by private insurance companies as well as government-sponsored health insurance. All are concerned with cutting skyrocketing medical costs. Yet, private companies have historically provided little insurance support for alternative therapies that are less costly and potentially efficacious. There is enough evidence of efficacy for some alternative cancer treatments to warrant preliminary support from insurance companies, pending outcomes analyses. A standard, three-week treatment in most of the Tijuana cancer clinics in 1995 ran about five thousand dollars per week; this is less expensive than a standard course of chemotherapy and radiation for many patients. Much less controversial are the various exercise, diet, and stress reduction plans for heart disease. In a few cases some private insurance companies have already realized the benefits of allowing patients to use some alternative therapies that meet profit criteria of lower costs and some efficacy. These criteria are usually met in cases of chronic disease where patients and companies are not happy with success rates and costs of standard treatments (Cowley 1995).

By the mid-1990s competitive forces were driving some health maintenance organizations into offering plans that provide various types of alternative medical coverage, and it is likely that market forces will continue to force the expansion of coverage of alternative therapies. Legislators could speed this process along and simultaneously lower health care costs by building in an option for patients on government-sponsored health insurance to pursue at least some alternative therapies that are safe, of equivalent or lower cost, and potentially efficacious. By leading the way, government-sponsored plans and outcomes analyses would provide a model for private insurance companies to follow, and, in the process, costs of the federal and state programs could be slashed considerably without encountering taxpayer anger. Clearly, these changes will not involve blanket acceptance of all alternative therapies. Specific decisions would have to be made pending cost-benefit analyses that take into account safety, cost, and outcome in comparison with standard therapies for the particular disease. However, there is evidence that at least some of the alternative cancer-therapy protocols have equivalent mean survival time and lower cost than conventional therapies (e.g., Hildenbrand et al. 1995). It is likely that these patterns would extend to other chronic diseases, particularly arthritis, heart disease, and AIDS.

Rights to Research Information

A more complicated area of policy intervention involves medical research priorities. In addition to providing the public with greater access to a wider range of medical-treatment options, a reformed medical policy also needs to provide the public with better guidance on efficacy. Statistics for 1990 indicate that the American people spent over ten billion dollars for alternative medical care, roughly equivalent to their out-of-pocket expense for hospital bills (Eisenberg et al. 1993). It is reasonable to assume that they would shift more financial resources to alternative medical care if they had more information and more insurance support. If they had more information, they would probably also spend the money more wisely by making better decisions about relative efficacy among alternative therapies. Given the fact that the American people are spending money on alternative medical care, and lots of it—with or without insurance support, and with or without the sanction of the medical profession or government agencies—the key question seems to be how to provide us with information so that we can make decisions wisely.

One problem in getting the accurate information to the public is the revolving door of funding among the NCI, the ACS, and the major cancer research and treatment centers. Economists James Bennett and Thomas DiLorenzo have shown how the research institutions represented by the American Cancer Society's board of directors also receive a large portion of the grants given by the organization.[8] Following alternative-cancer-therapy researcher Ralph Moss (1989), they point to the interlock of ACS and NCI personnel on many committees. Bennett and DiLorenzo also note that even though the ACS budget is much smaller than that of the NCI, "the ACS tail is wagging the NCI dog." (1994: 170). In general, the peer-review system tends to preclude innovators and to maintain the circulation of grants and research achievements among a relatively small group of people. Bennett and DiLorenzo advocate sunshine laws for large charities as a step toward opening up the entire process of research and education to public accountability. Their proposal is an important piece of the puzzle because the public needs to have more access to information on how research and funding agendas are being set. However, more reforms are also necessary.

A more direct form of intervention in agenda-setting was the Congressionally mandated creation of the Office of Alternative Medicine within the Office of the Director of the National Institutes of Health. Notwithstanding the good intentions behind this move, the budget for 1996 was set at under ten million dollars out of a quarter of a billion dollars for the Office of the Director and twelve billion dollars for the NIH as a whole. The public is going to continue to use alternative therapies no matter what the NIH does. However, without guidance in the form of a large quantity of well-designed research, the public is forced to make its decisions based on anecdotal information, patient referrals, and best case reviews. When I toured the alternative-cancer-therapy clinics and hospitals in Tijuana, I was impressed by the lack of statistical information on outcomes. Some had a track record of thousands of patients, but they were unwilling or unable to provide potential patients/consumers with clear statistics. Of the clinics and hospitals that I visited, only the one affiliated with the Gerson Research Organization had a published, peer-reviewed, quantitative analysis of outcomes, and the Contreras hospital was the only other institution I visited that offered a written evaluation of outcomes. The public badly needs controlled, clinical trials that are designed to provide a fair test of the strengths and weaknesses of alternative cancer therapies. In addition, there needs to be an immediate analysis of the outcomes of alternative cancer treatment by the retrospective analysis of existing collections of patient records.

In order to achieve this level of evaluation, the national cancer policy would have to go well beyond the drop-in-the-bucket funding level for the Office of Alternative Medicine. A drastic overhaul is needed by which Congress mandates much more money for cancer prevention and for the evaluation of alternative therapies. Given the fact that the current spending of over two billion dollars a year has been correlated with increasing cancer rates and few improvements in survival rates, it is time to overhaul the budget completely and dedicate the bulk of it (say one-third each) to prevention and testing of alternative therapies (with the other third remaining to support ongoing conventional projects). One might argue that approximately one-third of the NCI budget is already dedicated to "prevention," but as historian Robert Proctor demonstrates much of the spending takes the form of "chemoprevention," that is, trials of the effectiveness of chemotherapy as a preventive measure (1995: 267). Thus, legislation needs to redefine prevention to include public education programs. Furthermore, legislation is also needed to reorient research priorities away from basic research toward clinical studies. A step in this direction is Senator Mark Hatfield's proposed Clinical Research Enhancement Act, which would modify the focus of the NIH toward clinical research proposals and potentially open doors toward testing alternative therapies. This kind of proposal needs to be made more specific for the NCI, so that at least a third of the budget is devoted to testing the major nutritional and metabolic therapies that the public is already using, with or without guidance from the cancer establishment.

Reallocation of a substantial part of the budget of the NCI for primary prevention and testing alternative therapies would be an important step toward ending the war on cancer, but another problem would have to be solved in order for the funding to be successful. The great problem with controlled, clinical studies of alternative cancer therapies is that there is a history of mistrust based on suspicions of bias. For example, when public pressure finally led to clinical trials of laetrile in 1980, cancer activists became mistrustful because their offer to provide laetrile was not accepted. To this day many believe that real laetrile was never tested.[9] Likewise, with the Vitamin C trials Linus Pauling found a number of major design flaws in both the first and second Mayo Clinic studies. The flaws included discontinuation of the vitamin at any sign of worsening (thus provoking a rebound effect), a departure from the original protocol by reducing median time of Vitamin C administration to ten weeks, and evidence that members of the control group were surreptitiously taking Vitamin C.[10] Similar problems and mistrust have surrounded the hydrazine-sulfate trials of the early 1990s.[11] A key issue is the choice of institutions as hosts for controlled trials. Regarding the subsequent, similar controversy over the antineoplaston trials, Moss commented:

> The NCI and OAM both failed to recognize that these institutions represented the enemies of alternative medicine. . . . We saw this with laetrile at Sloan Kettering, we saw this again with hydrazine sulfate, and we saw it with the Mayo Clinic in their flawed tests of Vitamin C. I am not surprised that Memorial didn't follow the Burzynski protocol. It was predictable.[12]

In the clinical trials of alternative cancer therapies, testing passes to major research centers, where design decisions are made without consultation with the original advocates. Changes in design are made, and then later it is announced that the trials fail. When the design changes become known, questions are raised about the intention behind the design changes. Given the long history of suppression, of which I have outlined only a small portion in this book, the questions of intent and charges of cover-up are not necessarily paranoid or misguided. The end result is that alternative medicine advocates reject the failed trials as biased, and the huge public expenditures for clinical trials only fuel the controversy rather than resolve it. The trials result in no meaningful separation of effective from ineffective therapies, but only a greater gap of mistrust between both sides. In short, there is polarization rather than approximation.

Can anything be done to avoid this problem? One possibility is to move public funding of multicenter trials into institutional locations where there is no evident bias against alternative medicine. Some of the OAM-funded projects represent a move in this direction. However, multicenter studies of alternative cancer therapies done outside the cancer establishment would probably provoke mistrust from that establishment. Thus, multicenter trials of alternative cancer therapies may need to include some established medical centers as well, although maybe not some of the more mistrusted institutions. The difference of whether an establishment or a nonestablishment site is used might even be analyzed statistically as one variable in the experimental design.

More important, alternative-cancer-therapy activists have suggested an additional way to remedy the problem of bias in gold-standard tests: legislative mandates would stipulate that qualified advocates of alternative therapies play a key role as coprincipal investigators at the design stage of clinical trials. If the advocates have the power to approve the design protocol prior to its initiation, they would be in a weak position to cry foul by making post hoc analyses of design flaws in the event of a failed trial. Conversely, if an individual site changed the protocol without the advice and consent of the coprincipal investigator, it would be much more evident and reprehensible. Involvement at the design stage could also prevent the intentional or unintentional introduction of artifacts. In short, both sides would benefit. Furthermore, the taxpayers would not waste their money on huge clinical studies that end up being mistrusted because of potential flaws in the design that only come to light after the protocols are revealed and the studies are completed.

To turn this idea into practice, Congress needs to mandate that multicenter university trials of alternative cancer therapies include 1) some sites where the institutions are known to be friendly to alternative medicine; and 2) coprincipal-investigator status for the major advocate of the alternative therapy that is being tested. The coprincipal investigator would have the right to approve protocols and protocol changes. In addition, it may be necessary to set up a protocol review board to provide for a mechanism of adjudication should disputes arise among coprincipal investigators. This board could be a separate entity or a function of the OAM advisory board.

The policymaking process could be further democratized by backing up any boards that oversee alternative medicine research with a second form of public input into the research process. As discussed by STS analyst Richard Sclove (1995, 1996) of the Loka Institute, several European countries are now following the Danish model of citizen-review panels for policy decisions in areas involving scientific and technical controversies. This proposal goes beyond the idea of a "reasonable center" of experts on an advisory board to a lay review board composed of ordinary citizens. These citizens—perhaps in this case cancer patients who have used alternative therapies—meet at government expense in a forum similar to that of a jury. Over a period of several months they review research and funding decisions on a particular therapy, and they issue a report with policy recommendations.

What kinds of alternative therapies should be tested in a reformed world in which there is a vastly expanded OAM budget with a clear legislative mandate to evaluate alternative cancer therapies? Consider for a moment why I have examined in this book the bacterial-etiology theory. Prior to this study I have spent years studying alternative sciences and medicines in Brazil and the United States. Most of what I saw I consider very unscientific. I would not call it quackery or fraud because, for example, the healers I studied in Brazil were generally religious people who sincerely believed in what they were doing. However, I remain doubtful of claims of efficacy for ailments other than mild psychological disorders that were often interpreted as spiritual ailments and for which the spiritual idiom sometimes served as a successful translation of psychological problems.

In contrast, the bacterial-etiology theory—at least in the modified form that I presented in the previous chapter—has to its credit a body of scientific and medical publications, many in peer-reviewed journals. To repeat, the theory as I have modified it does not claim that bacteria cause cancer, but only that bacteria play an overlooked etiological role in some human cancers and that recognizing this role might help researchers to make better decisions for the testing of bacterial vaccines, choriogonadotropin (CG) vaccines, and antibiotics in cancer therapy. As discussed in the previous chapter, the rapid changes of medical consensus on ulcers—and more recently, a possible change on arthritis—suggest that bacteria may have been too quickly rejected as etiological agents in chronic, degenerative disease (Marshall et al. 1988, 1994). It seems appropriate to dedicate more of the taxpayers' money to research on bacteria and chronic disease in general as well as the possible link with cancer in particular. Certainly experiments to test Koch's postulates with CG-like-producing bacteria would be in order, and additional evaluation of the Glover serum, Livingston vaccines, Coley's toxins, and other bacterial vaccines seem worth the public investment. By additional evaluation I mean both attempts to analyze how these therapies might work—either guided by the bacterial theory or by other theories such as the cascade of cytokines theory—and to test the extent to which they do in fact work in human populations. Furthermore, any research on CG, either guided by the bacterial theory or not, should be supported.

I have used the case of bacteria and cancer merely as an exemplar and a case study. It should be clear that although I think the bacterial theory warrants some limited taxpayer funding from the twelve-billion-dollar NIH budget (one percent might be a good starting point), this dedication of public funds should happen alongside a larger allocation of funds to evaluate a number of other alternative cancer therapies. The great problem in budgeting is that there are many alternative therapies and alternative medical theories in circulation, and it would be a ridiculous waste of taxpayer dollars to test them all. Some criteria are needed to separate the more credible from the less credible alternative therapies. In this book I have suggested a combination of consistency, accuracy, pragmatic/efficacy, and low-social-bias criteria as a means for evaluating an alternative theory and therapy.

To make the first cut in a huge world of alternative therapies, I suggest that a good way to distinguish therapies that are worth investigating from those that are not is a general accuracy and consistency criterion that I call the "credible biological mechanism" test, provided that it is combined with a pragmatic criterion of some clinical evidence for efficacy and nontoxicity. If advocates of a therapy for cancer can provide a rationale and a modicum of evidence to support a theory that a credible biological mechanism is involved, and if they can point to some clinical evidence that the therapy has some efficacy and safety in human populations, then the therapy is worth investigating. The mechanism does not have to be proven to be true (as in the bacterial-etiology theory and the proposed CG mechanism), and the efficacy does not have to be demonstrated in more than case histories, because the goal is to provide research money to evaluate the therapy. However, the proposed mechanism of action should be within the realm of possibility given current biomedical knowledge, and there should be at least some solid case histories to indicate some promise for clinical efficacy.

The theories and therapies that, in my opinion, would meet a sufficiently credible biological-mechanism test and preliminary clinical-efficacy test generally fall into the overlapping categories of immunological, nutritional, herbal, and metabolic. In the area of alternative cancer therapies, a partial list includes bacterial vaccines, some alternative pharmacologic substances used in immunotherapy (e.g., antineoplastons and immunoaugmentative therapy), large-dose dietary supplements (vitamins, minerals, amygdalin, enzymes, and fatty acids), adjuvant psychotherapies with measurable psychoneuroimmunological markers, the Gerson diet and related diets that can propose a credible biochemical rationale (such as therapeutic potassium-sodium balances), and some herbs that contain known antitumor factors and have a long clinical record. Advocates of these categories of therapies generally can provide a sufficiently specific and credible biological rationale, and they provide some case studies that point to potential clinical efficacy. (Some have gone even further to quantitative case reviews and outcomes analyses, as in the case of Coley's toxins and the Gerson therapy.) Therefore, in my opinion, these therapies warrant dedication of sufficient public funds for testing and evaluation, and the testing needs to be conducted according to the guidelines of fairness described above (such as coprincipal-investigator status, site selection to reduce bias, etc.). The proposed biological mechanisms of these therapies are not necessarily what the next century's scientists and patients will accept as true, but the advocates of these therapies can propose a rationale that is sufficiently credible to be within the realm of possibility. To bolster this point, it is worth noting that cancer researchers are already moving in this direction. Some of the nutritional therapies are likely to become integrated into conventional cancer treatment as valuable adjuvant therapies.[13] However, evaluation needs to consider their claims to efficacy as primary therapies as well.

According to this perspective, some alternative therapies would not warrant investigation at the present time, at least as preliminary candidates for funding. I would classify in this group those therapies for which a credible biological mechanism is absent. Examples include prayer and psychic healing, homeopathy, dietary/nutritional/metabolic approaches without a credible biological rationale or a record of good case studies demonstrating clinical efficacy, and therapies for which the rationale is a theory of subtle energies not in agreement with current physical knowledge (often non-Western in origin, such as Reiki therapy).[14] I would not label this group quack therapies; many of the advocates have noble intentions, do not make unwarranted claims, and have produced some cases of anecdotal, anomalous recoveries to support their claims. However, taxpayer dollars should not be expended based on vague promises when there are many other orphaned therapies that have a much stronger basis of scientific support. The therapies that lack such support should be relegated to the domain of private funding until they amass a large body of evidence for efficacy or they can propose a sufficiently credible biological mechanism. Given limited resources, taxpayer dollars should go to the categories of alternative therapies that can meet the standard of the credible biological-mechanism test and show some potential for clinical efficacy. This standard, like all legal standards, cannot be reduced to an algorithm. It will have to be articulated in general terms and elaborated over the course of time across specific cases that clarify the criteria via precedents (as in case law). Legislative mandate needs to encode this standard by specific reference to the types of therapies described above, in order to prevent government agencies from reinterpreting the test in such a way as to continue to exclude research on alternative cancer therapies.

Skeptics may oppose these proposals as a tremendous waste of taxpayer dollars, but this economic argument is misleading. If skeptics turn out to be right, and if the major alternative therapies mentioned above as passing a minimal pre-evaluation test all turn out to be inefficacious, then cancer victims and their friends will have better information that will help them to avoid making bad financial and health-care decisions. As a result, a potential drain on the economy much larger than the cost of testing the therapies will be ended with a relatively modest investment of taxpayer dollars.

Conclusions

By proposing a major legislative intervention into the cancer research agenda, I am assuming that research agendas have become too important to be left to the scientists and doctors. Although a degree of autonomy is always necessary and even desirable, and certainly the input of the experts is necessary, medical research is also about patients and their rights as citizens. Furthermore, biomedical research and its regulatory apparati are far from autonomous; rather, much of it is shaped by professional, pharmaceutical, and other financial interests. Thus, there is a legitimate public interest in democratizing the process for establishing general research agendas in many areas of science, technology, and medicine. This interest is especially pressing given the epidemic of cancer and the failure of current research agendas to stop the dying.

One way of shifting more control of the agenda-setting process to the public is by legislatively mandating specific research programs or agendas. A meager start occurred with the creation of the very poorly funded OAM. However, that process needs to be carried forward by a number of additional reforms:

provide sunshine laws for large public health charities (with large defined so that the laws themselves do not constitute a barrier to entry for the smaller, alternative nonprofit organizations such as the Cancer Control Society);

mandate a major change in public funding of cancer research so that a large portion of the budget is dedicated to the evaluation of nontoxic and alternative therapies and to meaningful prevention (not trials of preventive chemotherapy);

provide a mechanism such as coprincipal-investigator status so that advocates of alternative medicine can participate in the design of large trials and be able to overrule changes in design that introduce negative bias;

mandate multicenter trials of alternative therapies that include organizations known to be nonhostile to alternative therapies;

provide for a protocol review board (or a binding arbitration power for the advisory board) for disputes among coprincipal investigators over protocol;

mandate occasional citizen-review panels to serve as lay public juries that evaluate funding decisions on alternative medicine research, protocol review board decisions, and appointments to the review board(s);

specify criteria that prioritize alternative therapies that warrant taxpayer support versus those that do not (such as my proposed credible biological-mechanism test and clinical efficacy test based on case studies).

The changes that I am proposing can be achieved with little or no cost to taxpayers, because I am suggesting a way of reallocating existing medical research budgets rather than spending more money. Furthermore, it is likely that a significant reorientation in budgetary priorities for research on cancer (and perhaps other chronic diseases such as heart disease and arthritis) could result in significant long-term savings to the national economy and government health-care budget. Given the impending demographic explosion of baby boomers who will become senior citizens, and the continued increase in cancer rates, such a low-cost investment strategy could have a huge and beneficial payoff in future Medicare costs.

In addition to the specific set of recommendations for the research process, state and federal legislators need to debate a number of other major proposals that could significantly alter the shape of health care and rights to medical access in this country. Among the proposals that should be debated are the following:

allow all qualified health care practitioners to use so-called unconventional therapies provided that the condition is life-threatening or chronic, the treatment or procedure is not dangerous to the patient, and the practitioner discusses all options with the patient (a proposal now under consideration);

end the ability of federal regulatory agencies to raid the businesses, homes, and clinics of people who prescribe, use, manufacture, or sell medical or nutritional products associated with the holistic health movement, and restrict the regulatory powers to recalls of specific products that pose a clear and present danger to the public;

provide for the federal payment of court costs of the defendants of FDA raids if the costs exceed a set amount;

continue the classification of vitamins, minerals, herbs, amino acids, enzymes, and other food supplements as food rather than drugs, but allow the producers to make true health claims with substantiation from the scientific literature and require that producers undergo periodic audits from private companies to ensure that the nutritional agents claimed to be in the supplements are actually there in the proper quantity;

lower drug-and-device approval regulatory barriers according to the Jaffe principle that relates level of proof needed to establish efficacy of the drug or device to the product's safety, but that also lowers the level of evidence for efficacy required for chronic diseases with no known cure such as cancer and arthritis;

provide a fair appeal process for advocates of unconventional drugs and devices who believe they are facing obstruction in the FDA regulatory approval process;

provide penalties for regulatory officials who knowingly or prejudicially misrepresent advocates of alternative therapies or the therapies themselves or obstruct fair application of regulatory procedures;

reorient Medicare and Medicaid funding to allow reimbursement for alternative therapies judged to be of equivalent or better efficacy, safety, or cost to standard therapies.

The proposed changes would result in minimal or no increased costs to the government, and some of the proposals could result in significant cost savings. In general, however, there is an overall savings to the economy that is incurred by allowing consumers to have access to more information and more resources so that they can make better decisions.

The war on cancer has earned the nickname of the "medical Vietnam." Scientists have become lost and defeated in the jungles of basic research, and patients listen only to the hollow and receding promise that a cure is at hand. It is time to declare an end to the war on cancer (and to the masculine metaphors), to bring home the men and women from the research fronts that have led nowhere, and to put them to work on meaningful prevention and on testing the alternative therapies that the public is already using.

6

Appendix

The New Science Studies

In the past it has been common for social scientists to adopt a neutral stance toward competing claims in scientific controversies. The social scientist's role was to examine the social dynamics of controversies in a way that scrupulously avoided taking sides. The policy implications of such studies, to the extent that there were any, were largely limited to suggesting ways of improving processes of dispute resolution in science or in the government.

The position of the neutral, observing social scientist who analyzes controversies certainly has its place; however, I find the position inadequate in the world today. One reason is that the more marginalized group will tend to capture an apparently neutral social-science study; in other words, it will claim that the neutral study supports its claims (Scott, Richards, and Martin 1990; Martin 1996). Likewise, the more established group will tend to reject the same study as covertly favoring the more marginalized group. In short, apparently neutral social studies, like those of the natural sciences, are often unpacked and translated as socially interested. No matter how rigorously social scientists attempt to adopt and maintain a neutral position, their accounts end up getting repositioned as favoring one side or the other.

Moreover, during recent decades the role of the citizen in science and technology controversies, especially those involving biomedical research, has become much more sophisticated. As scientific controversies have become increasingly public affairs, citizens have grown confident enough to question expert opinion. Nowadays public-interest groups routinely question the expert opinions, either directly or through other experts whom they enroll as allies. Social scientists occupy a unique position shared by only a few other occupations (such as journalists, librarians, some elected and appointed public officials, and nongovernmental leaders) because they are citizens with relatively unusual resources: the research skills and time that make it possible to understand the technical details of controversies, the freedom not to worry about losing their medical licenses or NIH grants, and the research expertise that allows them to find relationships between apparently neutral technical positions and interested social ones.

The changing role of the social scientist with respect to technical controversies is more than a product of the failure of the neutral observer model and the increasingly public nature of scientific controversies. It is also a product of the changing political economy and demography of universities. As governments cut back funding for higher education and as employment-conscious students move toward technological and pragmatic fields, the role of the social sciences as part of a general liberal-arts education has increasingly become a luxury that many universities cannot afford. Social scientists find themselves in interdisciplinary research-and-teaching teams that study, for example, new information technologies, the environment, biomedicine, and engineering. As the technical experts make decisions that have political and social implications, and as they make claims about the social or policy aspects of science and technology, social scientists are drawn into controversies as active participants rather than as disinterested observers.

For these and other reasons, the position of impartiality in the social studies of science is no longer adequate. Social scientists can and should play a role in the evaluation of which research programs warrant public support and expert attention. Although the details of specific knowledge claims and research choices within a research program may require expert knowledge beyond the level of competence of outsider evaluators, the choice among general research programs themselves is a matter of great public interest. Social scientists, like other citizens, have a legitimate right to become involved; indeed, we have an obligation to exert a leadership role.

STS after the Strong Program

Readers familiar with science and technology studies (STS) will undoubtedly recognize that the framework I advocate here departs substantially from the basic principles of the strong program. Introduced in the 1970s, the strong program is largely recognized, for better or for worse, as a watershed in the social studies of science. Prior to the 1970s, most sociologists focused on institutional aspects of science such as its reward system or its networks of invisible colleges. During the 1970s some sociologists turned back to an older research tradition known as the sociology of knowledge.

Their work led to a new emphasis on social studies of the content of science, in other words, its theories, methods, and knowledge claims. The term "constructivism" has different meanings for different users, but here it will be interpreted as a range of frameworks that emerged in the wake of the strong program (see Hess 1996, 1997). These frameworks include laboratory studies of the social construction of knowledge and actor-network theories of scientific entrepreneurs.

The strong program represents a concise and widely cited early formulation of the sociology of scientific knowledge (SSK). The basic principles were: 1) causality: social studies of science would explain beliefs or states of knowledge; 2) impartiality: the sociology of scientific knowledge would be impartial with respect to truth or falsity, rationality or irrationality, or success or failure of knowledge (and, presumably, technology); 3) symmetry: the same types of cause would explain true and false beliefs, etc. (in other words, one would not explain "true" science by referring it to nature and "false" science by referring it to social interests or cultural values); 4) reflexivity: the same explanations that apply to science would also apply to the sociological studies of science (Bloor 1991).

For a number of reasons the strong program and its siblings and offspring in the sociology of scientific knowledge became exhausted. One problem was that SSK researchers tended to dismiss other currents of STS research: the institutional sociology of science, often called the "Mertonian" sociology of science, for its lack of interest in explaining content; critical social studies of science and technology for their lack of impartiality, and the traditional philosophy of science (and the view of many practicing scientists) because it often explained the outcome of scientific disputes by reference to the natural world instead of to the social world, a move that violated the symmetry principle. By far, the third problem was the most severe, and philosophers were soon levelling charges of relativism and incoherence against the strong program and constructivist accounts of science.

The main philosophical objections have been epistemological and moral relativism. The argument on epistemological relativism goes roughly as follows: the impartiality and symmetry principles imply that accounts of how knowledge came to be widely accepted in science should rely only on sociocultural factors, rather than on representational (evidentiary) factors. In other words, as a variable in explaining how knowledge came to be accepted or constructed, such as after the closure of controversies, SSK ignores the accuracy or consistency of truth claims. Some SSK researchers argued that the real, material world—and observations drawn from it—

plays little or no role in the resolution of controversies. Philosophers pointed out that social scientists who analyzed natural-science controversies went about resolving their own sociological controversies by using the evidence of social facts. In other words, in practice social scientists granted themselves a rationality that they were unwilling to grant scientists in their sociological accounts of how knowledge came to be widely accepted.

A second charge against SSK was that the extreme epistemological relativism leads to moral relativism. This charge tended to emerge more from the critical or engaged wing of STS. Clearly, anyone who wants to change the world, or at least to change aspects of science and technology, could not rely on a framework that endorsed epistemological relativism. To change the world one must first be able to justify one's diagnosis of what is wrong with it and to justify proposals for alternatives. STS researchers who were associated with the scientists for social-responsibility movement, the feminist movement, or the radical-science movement also tended to be skeptical of the epistemological relativism of the SSK school.

At first the relativism controversy remained within the STS field and it had a largely disciplinary and national flavor: philosophers versus sociologists, socially concerned scientists versus philosophically concerned social scientists, and American sociologists versus their British counterparts. However, by the 1990s the controversy had expanded into the so-called "science wars," when some scientists began to take the few radical relativists of the SSK school as representative of STS in order to discredit the field as a whole. This misrepresentation is unfortunate and highly polemical. As has already been indicated, the STS field is quite diverse and characterized by fundamental philosophical disagreements.

The entire relativism controversy is largely a dead horse that some philosophers and STS-bashers managed to keep alive into the 1990s. They often propose an alternative of naive scientific realism, which on this issue would hold that in most historical cases mere research alone (such as experimental data from a crucial experiment) has resolved a scientific or medical controversy. This alternative is untenable as an empirical description of science because it simply does not match the historical record. Nevertheless, the philosophers and other critics of SSK relativism are right in rejecting the position of extreme epistemological relativism. I am among a large number of STS analysts who support neither the extreme relativism advocated by some SSK researchers nor the naive realism of some of their detractors. Rather, there is a reasonable middle ground. From this viewpoint social contingency and cultural values shape the process of knowledge-making in tandem with the restraints of evidence, that is, the structures and limitations of the real, material world. Some philosophers have developed versions of this position under the rubric of "constructive realism" or "realistic constructivism" (e.g., Giere 1993, 1995; Fuller 1993: 5). The question of how much the social/cultural shapes the outcomes of controversies, versus how much considerations of evidence do, cannot be answered in general terms: each case is different. Certainly, in the case of cancer research—a science that is applied, industrialized, and supported by a large system of private and public grants—social/cultural factors will loom large. However, considerations of evidence—including points in the history of cancer research where the evidence was ambiguous—are crucial for the account as well. The point is to develop a sophisticated analysis that falls prey to neither a naive relativism nor a naive realism.

I also part with the strong program and constructivist accounts of science by moving from a descriptive account of a controversy to a prescriptive account. I ask, for example, what criteria should be used to evaluate the bacterial-etiology theory and research program? Is it credible science? Should taxpayer dollars be spent to investigate alternative cancer theories and therapies? If so, how should government institutions be reformed in order to make such investigation more viable? The questions about the value of competing research programs are fundamental to any scientific field, and social scientists need to ask those questions if their work is to be relevant to public debate and not an exercise in academic parochialism. Once one begins asking that type of question, relativism is impossible. Likewise, a naive faith that evidence and research will solve all problems is a recipe for policy failure.

This book is based on a set of principles that provides an alternative to the strong program and relativist versions of constructivism. The principles provide one translation of a constructive realist position into a concrete, interdisciplinary research program for STS. Growing numbers of STS analysts already use some or all of these principles as part of their research frameworks. In effect, these principles could be interpreted as a distillation of the major lines of postconstructivist STS. The principles are as follows: 1) power: the analysis focuses on how scientists in positions of power shape the constitution of orthodoxy and heterodoxy, of consensus and controversy, in fields of science and technology; 2) culture: the interpretation of the history of power in science and technology must go beyond instrumental reason to include an analysis of the cultures in which researchers are working, including the role of evidence in controversies; 3) evaluation: the STS analyst should analyze the credibility of competing scientific claims of alternative positions by using the best available knowledge and by taking into account the possible biases in that knowledge uncovered by previous steps of the analysis; 4) positioning: the general political and specific policy implications of the research should be examined and evaluated. This approach to STS research involves a significant shift from the neutral, unengaged position that characterizes many studies in the field, and it is sympathetic to the alternative STS traditions associated with some versions of feminism and the various science-for-social-responsibility movements.

To put into practice this framework, it has been necessary to draw on a wide range of STS disciplines. Unlike the SSK researchers, who tended to exclude significant branches of STS research as irrelevant or wrong-headed, I draw eclectically on various methods and concepts in a more transdisciplinary spirit. As a theoretical contribution to the STS field, this framework provides one way of moving beyond the disciplinary ghettos to a transdisciplinary analysis in which history, sociology, anthropology, cultural studies, philosophy, and policy studies all have a role to play in a synthetic analysis.

Power and Culture

The first principle is that the analysis is political in the sense that it pays attention to the operation of power in the history of what comes to be accepted as scientific knowledge and what comes to be rejected as unorthodox or heterodox. The term "power" has many meanings, among them the Marxist sense of the arrangement of society so that one class dominates and benefits, the Weberian sense of one group being able to get its way even when other groups want something else (played out here in the form of professional power), the Foucauldian sense of power embedded in practices that rationalize and discipline social life, and the feminist sense of power as the differential effects of social arrangements on genders and other social categories. The first level of analysis of power addressed in this book is a very basic form of power: the suppression of dissident traditions in science and medicine. However, beyond the sometimes graphic and brutal means that those in power use to maintain their hegemony, my concern is also with more complex forms of power that require embedding an analysis of power in an analysis of culture.

The question of heterodoxy is a good empirical topic for developing a cultural approach to the analysis of power. Whereas many SSK accounts focus on how successful science and technologies achieved success, I focus on how a research program came to be forgotten and excluded. Although some SSK frameworks examine the failure of scientific networks, the accounts focus on strategic errors and technical bottlenecks. From my perspective, this approach is too narrow. The question is not merely why some scientists did not became the Pasteurs of their field; rather, it is also why the field did not allow them to achieve success.

The existence of heterodoxies provides one reason why an analysis of power needs to be accompanied by a more general analysis of culture. It is helpful to begin by returning to one of the first research programs associated with the strong program: Edinburgh-school interests theory. Although I do not support a simplistic return to this framework, I am interested in the history of that originary moment in SSK when certain theoretical assumptions were excluded, much as in the originary moments of cancer research in the early twentieth century when microbial theories of cancer were foreclosed. The case studies of the "Edinburgh school" contributed to STS a model of how to analyze scientific controversies in terms of interests (e.g., Barnes and Shapin 1979; MacKenzie 1983). A variety of interests were considered, but class interests were the most controversial, and criticisms of the Edinburgh school tended to focus on its attempts to draw lines of causality from class interests to technical positions in a scientific controversy. As various critics argued, the interests studies suffered from the problem of transforming macrosociological class interests into the microsociological accounts of the motivations that shape the action of individuals. In general, interest-based analyses risked turning scientists and other technical actors into "interest dopes." The framework reduced analyses of scientists to flat, cartoonlike characters who were guided by exogenous interests rather than a complex set of contingencies and motivations, including a concern with evidence and consistency. Given the general overdetermination of social action, unicausal theories of any sort, including some sort of determination by class interests, are now highly suspect to many STS analysts (Barnes 1981; MacKenzie 1981, 1984; Woolgar 1981a, 1981b; and Yearley 1982).

This version of interest theory, which I discuss in chapter three as Version I, runs into exactly the same problems when I attempt to apply it to the case of cancer research. In other words, it is very difficult to draw any causal lines of influence from the capitalist class to the content of cancer research. The problem emerges even though medical research in the United States should provide good case-study material because of the potential vector of transmission provided by the capitalist foundations that supported the transformation of medicine during the early twentieth century. The basic imputation problem remains, and as I examined the historical record closely, the lines of causality dissolved into ambiguities of interpretation.

Why, then, bring back the dead horse of Edinburgh-school interest theory? When interest theory was rejected, most forms of macrostructural sociological and anthropological analysis were also rejected in the SSK wing of STS. There is still a prominent role for interests analysis, at least in the case of cancer research, as is amply demonstrated by the huge financial interests that are at stake in the conservative consensus of the cancer research community. Thus, professional and financial interests play a major role in the constitution of this research tradition, but not in the form of determination by capitalist class interests. Rather, the interests are more local: they involve the emerging cancer industry. In fact, it is difficult to provide a good explanation of the history of cancer research without referring to the substantial institutional and financial interests of what some estimate has become a one-hundred-billion-dollar-a-year industry in the United States alone.

An analysis that is restricted to professional interests still represents a very limited form of power, similar to what Marshall Sahlins (1976) calls practical reason. Practical reason continued to inform many SSK accounts of scientists even after the Edinburgh-school program was rejected (e.g., Latour 1987, 1988). Actor-network theory alternatives to interest theory portrayed scientists as acting out of a sense of strategic gain rather than a desire to defend professional or other interests. The primary metaphor that emerges from these studies is military: scientists are like generals on a field of battle where credit, credibility, fame, and glory are the stakes of victory. The warfare metaphor sometimes overlaps with the economic metaphor that also appears in some SSK accounts: both are stories of actors who are attempting to maximize gain. Scientists become mere strategists; instead of being driven by interests, they attempt to produce interests and enroll allies in an effort to further their schemes of self-aggrandizement or enhance their credibility.

Just as it is inadequate to explain the history of repression and suppression as the result of a coalition of interested doctors, researchers, and institutions that act in a strategic way to protect their interests, it is also inadequate to explain that history as motivated merely by a masculine, military quest for power and credit. Additional analytical resources are necessary to explain why consensus knowledge developed as it did. Those resources need to incorporate another level of analysis than the instrumental reason that underlies a variety of SSK frameworks, including both Edinburgh-school analyses and actor-network analyses. The rubric of "culture" flags this other level that takes a sociology of science beyond instrumental reason. Scientists act inside research cultures, which are in turn nested in more general societal and temporal cultures. The analysis needs to examine the general parameters of these research cultures—their key assumptions and polarities of orthodoxy and heterodoxy—and their relationship to the general societal and temporal cultures.

Bringing culture into the analysis of power makes it possible to move beyond the instrumentalism of some SSK accounts without losing the original insight of the Edinburgh-school studies: that science is made in societies and not merely in networks. As chapter three demonstrates, the failure of filtrationism as a school of bacteriology and the hegemony of cytotoxic therapies cannot be explained merely as the failure of the microbial researchers to build strong networks or make sure that powerful financial interests were appeased. The analyst also needs to enter into the culture of medicine and science to understand why some possibilities were marginalized and why others were easy to incorporate, in short, why the playing field was not even.

Describing the culture of research programs bypasses the simplifying accounts of scientists that reduce them to entrepreneurs, military strategists, interest dopes, or other figures of practical reason. Instead of the two-dimensional flatland of practical reason, cultural analysis delineates the key theoretical and evidentiary claims of the research cultures in which scientists worked. Thus, the analysis restores some of the rationality of science that many scientists find missing when they read SSK accounts of their fields. At the same time, this framework does not return to a merely internalist account that focuses only on theories, evidence, and other so-called rational factors in the resolution of controversies and the making of consensus. Although the rationality of science is restored by examining the scientific world and assumptions of the scientists, the culture of research programs itself becomes the object of social scientific inquiry.

As in the case of the relationship between culture and ecology in the anthropological analysis of small-scale societies, the relationship between the specific research culture and the ecology of the broader society is the focal point of social inquiry. For example, in the case of twentieth-century cancer research in the United States, a modernist culture that emphasized standardization, a masculinist culture that emphasized heroic medicine, and a relatively insulated national culture that precluded widespread international travel and communication are among the additional factors that helped structure cancer research so that the general microbial theories fell off the playing field. Conversely, the postmodern culture of flexibility, the emergence of feminism and its linkage to breast-cancer activism, and the globalization of communication and concomitant de facto deregulation are among the factors that are spurring the growth of alternative cancer therapies today, including a re-examination of the lost bacteria-and-cancer research program.

Although macrosociological interests and broader cultural values do play a role (along with evidentiary and consistency concerns) in shaping the research program's culture, their shaping role is mediated through that culture, which also mediates the changing forces of new empirical evidence. Like a prism, research cultures do not always reflect the external factors in a simple way. As Marshall Sahlins (1976) argued for small-scale societies, cultures develop their own internal logic that changes in response to outside shaping interests and historical contingencies, but these changes tend to occur along lines that are consistent with the cultural logic. Likewise, scientific research programs change over time (in response to new research findings, the growth of new industries, changing professional configurations, etc.), but they do so in ways that tend to be consistent with cultural logics that have a momentum of their own. In this way, power, like knowledge, is mediated through culture, and the framework makes possible a more complex understanding of power and culture in science.

Evaluation

An analysis that moves beyond instrumental reason to a cultural level nevertheless remains within the parameters of descriptive and explanatory social science. One could finish reading such a study and still have no idea whether, for example, the microbial theories of cancer have any scientific value. "So what?" one might legitimately ask. Remember, the question of credibility is not a theoretical exercise for people with cancer.

Another step of analysis is required: to examine the comparative validity of the knowledge claims of the consensus and alternative research traditions. In other words, fully interdisciplinary STS analysis must step out of the traditional plane of social scientific analysis (here formulated around the two strands of culture and power) to a prescriptive level. This level involves two steps: the evaluation of knowledge claims and the evaluation of proposed policy or political changes.

The evaluation of knowledge claims is based on the standards of the best available scientific knowledge at the time of the evaluator's analysis, but it also assumes that those standards may themselves be biased against the research under analysis due to the same political/cultural processes already analyzed. Far from providing an impartial or symmetrical analysis, I go on to evaluate the research claims. Yet, I do this by taking into account the possibilities of bias introduced by the cultural politics analyzed in the previous stages.

The evaluation of research programs generally cannot be reduced to a formal method (such as a crucial experiment); rather, it is more like the processes of dispute resolution in the legal profession and the qualitative social sciences. Evidence can be established but always within a context that recognizes the power of cross-examination and interpretation. To establish criteria for evaluating the alternative research program, I draw on a wide range of sources in the philosophy of science. Rather than viewing the philosophy of science as a straw person, as often occurs in the SSK literature, I view it as providing a basic set of resources that are fundamental for the evaluation problem. As philosopher Steve Fuller (1988) has argued, philosophy of science can be relevant for a wide range of public debates when it is geared toward helping people clarify their positions on prescriptive issues. I focus on one specific prescriptive problem, theory choice and evaluation.

Thomas Kuhn's list of prescriptive theory-choice criteria is a good starting point because it brings together the findings of a number of philosophical traditions. Although Kuhn has sometimes been accused of radical relativism, his essays subsequent to the *Structure of Scientific Revolutions* reveal that the accusation is unfounded. In one of his subsequent essays, Kuhn outlined the major criteria that should guide sound theory choice:

> First, a theory should be accurate: within its domain, that is, consequences deducible from a theory should be in demonstrated agreement with the results of existing experiments and observations. Second, a theory should be consistent, not only internally or with itself, but also with other currently accepted theories applicable to related aspects of nature. Third, it should have broad scope: in particular, a theory's consequences should extend far beyond the particular observations, laws, or subtheories it was initially designed to explain. Fourth, and closely related, it should be simple, bringing order to phenomena that in its absence would be individually isolated and, as a set,

> confused. Fifth—a somewhat less standard item, but one of special importance to actual scientific decisions—a theory should be fruitful of new research findings: it should, that is, disclose new phenomena or previously unnoted relationships among those already known. (Kuhn 1977: 321–22)

Kuhn's decision to begin with the accuracy and consistency criteria is well-founded. The accuracy criterion could be reconciled with some type of verificationism or falsificationism. Likewise, the consistency criterion is similar to the theory choice criterion favored by the conventionalist Pierre Duhem (1982). When one examines philosophers on the practical question of theory-choice criteria, there is little disagreement over the fundamental importance of accuracy and consistency as theory-choice criteria. Of course, as some philosophers have warned, consistency can have conservative implications, and therefore a highly inconsistent theory will have to do very well on other grounds.

Most of Kuhn's other criteria can be interpreted as corollaries of accuracy and consistency. Simplicity was also advocated by both the positivist Rudolf Carnap (1995) and the falsificationist Karl Popper (1963). Simplicity, however, must be judged against a background of other theories, and therefore ultimately I view it is a subcriterion of consistency. For similar reasons I do not include scope as a major criterion; it too must be judged against the background of other theories and may be considered a subcriterion of consistency or accuracy. Finally, I interpret Kuhn's fruitfulness criterion as a corollary of accuracy that is oriented toward future research.

Feminist philosophers have introduced a number of criticisms of standard theory-choice evaluation criteria, and Helen Longino (1994) distilled an alternative list of six criteria for theory (or research program) choice that are implicit in feminist STS analysts such as Sandra Harding (1986, 1992) and Donna Haraway (1991). Longino's criteria are empirical adequacy, novelty, ontological heterogeneity, complexity of relationship, applicability to current human needs, and diffusion of power. Clearly, empirical adequacy is more or less the same as the fundamental accuracy criterion for which there is a great deal of consensus. Novelty is a reminder of the conservative implications of over-reliance on consistency. Applicability to current human needs is an important development of the pragmatist tradition, and it deserves to be added as an additional criterion. In the case of alternative cancer therapies, this is a criterion of overwhelming importance for deciding which research programs should be funded. The operationalized version of the criterion would imply answering the question of how likely the theory is to lead to progress toward safer and more efficacious treatments.

Longino's other criteria introduce unnecessary ambiguities (Hess 1997). In their place I have opted for lower social bias such as gender or race bias as a significant theory-choice criterion that represents the general feminist contribution to philosophical discussions. For example, if one of the candidate theories appears to be more aligned with masculinist values and practices (such as a cancer theory that is aligned with the practice of what many view to be unnecessarily aggressive breast surgery), then this negative criterion should be relevant to theory choice. Alone, the criterion may not lead to rejection of a theory, but it can serve as a warning flag to re-examine a theory on other grounds.

In sum, I suggest accuracy, consistency, pragmatic value, and lower social bias as the four groups of major criteria for evaluating major theories and research programs. Clearly, it is possible that contradictions may emerge among the different criteria. In this case, one must use the criteria to triangulate (or quandragulate) one's way through what is obviously a complex decision that cannot be reduced to an algorithm. However, in my opinion accuracy should remain the most important criterion.

Positioning

The fourth guiding principle is that the analysis is positioned; it provides an evaluation of alternative policy and political goals that could result in beneficial institutional and research program changes. As a social scientist I therefore assume that I will be positioned inside the controversy, as the capturing literature demonstrates is inescapable, and that I am better off positioning myself rather than having someone else do it for me. In the terminology of the STS field, this level of analysis can be described as a type of reflexivity, but one that is more profoundly sociological or anthropological than previously discussed forms.

Elsewhere I argued against the mere epistemological reflexivity of the strong program and suggested a culturally oriented reflexivity that operated at the level of the relations between the researcher's academic communities and the groups researched (Hess 1993). The final chapter in the present book develops this transition further toward what might be best termed positioned intervention. The topic has received increasing attention in anthropological studies of science and technology (e.g., Downey, Dumit, and Traweek 1997). I develop the discussion of positioning by viewing it as intervention in a controversy by evaluating agendas for potentially beneficial institutional changes and investments for future research.

Although the final chapter examines policy issues, it also represents an attempt to discuss policy in a different vein. Informed by the SSK lesson that draws attention to content, I examine the issue of research policy. In other words, a crucial policy issue for national health is making a sound choice among competing research programs. It is not a question merely of developing more efficient ways to spend money on research, or of spending more money for research on a particular disease. Rather, it is a question of reorienting the research agenda toward prevention and environmental carcinogenesis, as Samuel Epstein and colleagues (1992) argue, as well as toward the evaluation of alternative therapies. In suggesting policy reforms, one has to take into account that there is a recalcitrant cancer establishment that has a documented record of dragging its feet and suppressing research on alternatives. Thus, policy needs to be rethought in terms of how to make sure that legislated intentions are carried out. One needs to think through the landscape of the research cultures to derive policies that will carry out the intentions of the legislators. In this way, the policy discussion draws on the insights of the previous stages of the analysis. Rather than view science studies as a series of disciplinary ghettoes—history, social studies, philosophy, and policy—I bring these fields together to provide a transdisciplinary approach to a pressing social and medical problem. The final stage, then, is a clarification of policy options that rests on transdisciplinary empirical research and philosophical analysis.

Notes

NOTES TO CHAPTER 1

1. On cancer incidence and mortality rate increases, see Davis and Hoel (1992); Davis, Dinse, and Hoel (1994a: 431) citing Ries, Hankey, and Edwards (1990); Epstein (1993); and Wing, Tong, and Bolden (1995: 12–16). Rates are also published by the American Cancer Society (1992) and National Cancer Institute (1991). Only a few cancers (such as stomach cancer) show a decline. The explanations for the growth in cancer incidence are the subject of heated debate (Proctor 1995: chs. 3–6; Wright 1991).

2. McGinness (1991) reviews the different estimates of users of alternative cancer therapies, and Eisenberg et al. (1993) provide the estimate of the larger population of alternative medicine users. For an overview of legislative changes, see Sale (1995). For reviews of major alternative cancer therapies, see Falcone (1994); Moss (1992); Pelton and Overholser (1994); and Walters (1993).

3. For sources on the controversy over survival rates, see Epstein (1993: 18–19); Proctor (1995: 4, 252); Lerner (1994: 51–54); Moss (1992, 1995); and Wing, Tong, and Bolden (1995).

4. See, for example, Lakatos (1978) and Laudan (1977), who in turn were developing a more specific alternative to Kuhn's (1970) concept of paradigms.

NOTES TO CHAPTER 2

1. Personal correspondence, Nauts to Hess, December 20, 1995.

2. Personal correspondence, Nauts to Hess, December 20, 1995, and March 29, 1996.

3. Personal correspondence, Nauts to Hess, December 20, 1995.

4. The first version (Parke Davis IX, 1899–1906) was "very weak"; the second (Parke Davis XII, 1906–15) was "a little more effective"; and the third (Parke Davis XIII, 1915–51), was "better than XII." Personal correspondence, Nauts to Hess, December 20, 1995. On Lister Institute products, personal correspondence, Nauts to Hess, March 29, 1996.

5. Löwy (1993: 340–41). The studies included Shwartzman and Michailovsky (1932) and Duran-Reynals (1933), followed by Shear et al. (1943).

6. In 1943 Murray J. Shear, at the National Cancer Institute, isolated what he

believed to be the active substance in the Coley toxins, a lipopolysaccaride in *Serratia marcenscens.* This discovery was "important to the later discovery of tumor necrosis factor" (Old 1988: 60). Old (1988) reviews subsequent research on TNF and immunotherapies, and Rook (1992) and Starnes (1992) discuss the revived interest in Coley's toxins. On the cascade of cytokines, see Wiemann and Starnes (1994). Research on TNF and bacterial vaccines is also mentioned in the Office of Technology Assessment report on BCG by historian Patricia Spain Ward (1996). According to Moss (1989: 128) and Hildenbrand in his foreword to Haught (1991: ix–xi), the OTA staff refused to circulate Ward's papers, including her history of BCG which mentions the TNF research. These papers became part of a bitter controversy and charges that OTA staff was manipulating contract reports for the final report (Office of Technology Assessment 1990; Moss 1996).

7. Personal correspondence, Nauts to Hess, December 20, 1995.

8. Personal correspondence, Nauts to Hess, December 20, 1995.

9. For example, *New York Post,* April 17, 1936; *New York Times,* April 16 and 17, 1936.

10. Nauts (1995: 25), amended by personal correspondence, Nauts to Hess, March 29, 1996.

11. Letter from Rhoads to John D. Rockefeller, Jr., December 29, 1941, Friends and Services, Doctors, William B. Coley, II 2 H, 26, 198, 231 C71, Rockefeller Archives.

12. For example, Havas, Groesbeck, and Donnelly (1958); Havas, Donnelly, and Levine (1960); and Johnston (1962).

13. Personal correspondence, Nauts to Hess, March 29, 1996.

14. Personal correspondence, Nauts to Hess, March 29, 1996. Similar issues plagued other trials of so-called unproven therapies. See Markle and Peterson (1980, 1987); Moss (1989); Pauling (1993); Peterson and Markle (1979a, 1979b); and Richards (1981).

15. Personal correspondence, Nauts to Hess, March 29, 1996. On clinical trials, see Havas et al. (1993) and Tang et al. (1991), and the review by Wiemann and Starnes (1994).

16. Nauts commented on this sentence as follows: "There is nothing mild about the intellectual repression Coley endured." Personal correspondence, Nauts to Hess, March 29, 1996.

17. See *Journal of the American Medical Association,* letters of January 1, p. 52; Feb. 5, p. 396; and March 26, p. 885.

18. McArthur to Coley, November 22, 1920, Institute for Cancer Research. The letters cited in this section were made available thanks to Helen Coley Nauts. In some cases I received hand copies of the letters because they were too frail to be copied mechanically.

19. Douglas to Coley, March 21, 1923, Institute for Cancer Research.

20. Glover to Coley, June 15, 1924, and Nerry (? illegible) to Coley, March 23, 1923, Institute for Cancer Research.

21. Mayo to Coley, July 3, 1926, Institute for Cancer Research.

22. Glover to Coley, March 25, 1931, Institute for Cancer Research.

23. Coley to Glover, March 26, 1931, Institute for Cancer Research.

24. Coley to Glover, February 2, 1932, Institute for Cancer Research. Berg also wrote Coley to confirm that he had independently cultured the organism: Berg to Coley, March 16, 1932, Institute for Cancer Research. See Berg and Coley (1932).

25. Coley to Mayo, April 5, 1932, Institute for Cancer Research.

26. Gye to Coley, March 17, 1936, Institute for Cancer Research.

27. Nauts to Hess, January 18, 1996.

28. "Dr. Netterberg spent many months tracking down two cancer patients whom Dr. Scott had cured of advanced cancer and whose cases were written up in the *Irish Journal of Cancer* in 1926. After being declared terminally ill with cancer, both had lived into old age and were completely free of cancer decades later" (Netterberg and Taylor 1981: 10). Scott also published in *Northwest Medicine* during this period. In 1977 Netterberg tried to get copies of Glover's NIH records directly from the NIH, and he was told that they were lost (Netterberg and Taylor 1981: 38). RLIN searches on Scott and Glover did not result in success, so I have relied on the book by Scott's friend Mark Boesch (1960).

29. Deaken to Murdock, June 24, 1926, Institute for Cancer Research.

30. Probably with the aid of Nikola Tesla, Lakhovsky developed a frequency machine similar to the one that Rife developed (Lakhovsky 1988). Work on electronic frequency machines in France continued after Lakhovsky with the controversial research of Antoine Priore (Graille 1984).

31. Unfortunately, Lynes does not document these final events surrounding the demise of the Rife machines; such documentation would be an important contribution to the historical record, and until they are fully documented they are subject to warranted skepticism. I attempted to get in touch with Rife's partner John Crane, but he did not respond to my inquiries. I later learned that he died during the winter of 1996.

32. Their therapy should therefore be classified as a general immunotherapy rather than a serum or vaccine for a proposed cancer-causing microorganism. Because the formula for Krebiozen was secret, it is difficult to assess the claims or the proposed biological mechanism. See Ivy (1956) for the research, the Council on Pharmacy and Chemistry (1951) and Stoddard (1955) for critiques, Bailey (1958) on the politics and repression of the therapy, and Ward (1984) for a professional history of the incident.

33. The biographical materials are from Livingston (1972, 1984).

34. Bisset (1969, 1970). A very short notice also appeared in the July 8, 1969, issue of the *Journal of the American Medical Association*.

35. See Crofton (1936); Young (1925a, 1925b); and Boesch (1960) for an overview of the battles faced by Young in the United Kingdom. Also not discussed here is the microbial research on cancer in Italy, such as the work of Fonti (1958), and the viral theories propounded in Spain by Duran-Reynals (1950) and in Japan by Hasumi (1980). Furthermore, while I was in Brazil in 1995 I found some evidence of a historical interest in bacteria and cancer there as well, including among well-known medical figures such as Miguel Couto.

Another prominent German case is that of Wilhelm Reich, who had to flee Germany because of his criticisms of fascism, and then later ended up in prison in the United States because of his unusual scientific experiments. Reich believed that pleomorphic organisms, which he termed "bions," played a role in cancer. Given the problematic relationships between the microbial cancer researchers and fascism in Germany, it would be interesting to explore the Reich saga in more detail.

36. Bowker and Latour (1987: 739–40) argue that the idea that the French educational system is highly centralized is a myth, but they agree that most research is funded by the government and is centralized through the national research organization CNRS. Freudenthal (1990) argues that the centralized funding of the research in France can explain the uneven success and dispersed institutionalization of science and technology studies in France in contrast with the Anglophone world. Thus, these studies confirm but also qualify Villequez's use of the well-known claim that French sciences can be conservative due to institutional reasons such as centralized funding and the way French society is centered on Paris.

37. According to Villequez (1955), this is reported in the *Semaine des Hôpitaux* (Supplément d'informations, June 14 and 26, 1965). Another of Naessens's anticancer drugs was GN-24, which was based on the theory that cancer cells could be stopped by a drug that blocked anaerobic respiration.

NOTES TO CHAPTER 3

1. Barnes (1977, 1981); MacKenzie (1981, 1984); Woolgar (1981a, 1981b); and Yearley (1982). See Martin (1993) for the social context that might help explain the abandonment of interests theory.

2. See Proctor (1995: ch. 2) on the history of the environmentalist thesis. The work of Maud Slye at the University of Chicago on hereditary factors in cancer was recognized at the Lake Mohonk conference (Coley 1926). Seibert (1968) mentions the influence of Slye's work, and Blumenthal (1991: 263) also mentions as important the related research of Clara Lynch of the Rockefeller Institute.

3. Ellerman and Bang (1908); Fujinami and Inamoto (1914); and Rous (1910).

4. Coulter (1987: 449), citing *Journal of the American Institute of Homeopathy,* V (1912–1913), p. 509.

5. Blumenthal (1991: 257); Fleming (1987: 208); and Flexner (1987).

6. Archival sources also indicate that John D. Rockefeller, Jr., and William Coley were friends, and a note indicates that Rockefeller, Sr., and a nurse respected Coley (Rockefeller Archives, Friends and Services, Doctors, William B. Coley, 231 C71, II 2 H, 26, 187). Laurence Rockefeller also gave a keynote address at the 1975 meeting of the Cancer Research Institute founded by Helen Coley Nauts.

7. Berliner (1985: 139–75); Corner (1964: 158–59); and Brown (1979: 166–67).

8. See Borkin (1978). Claims of the role of the cartel in the Nazi rise to power have probably been overstated (Stokes 1988: 22).

9. Rhoads papers, Rockefeller Archives, RG 608, Series 1, Box 2, Folder 18, and Series 4, Box 8, Folder 97; RG 1.1, Series 200, Box 98, Folder 1189. Also see Corner (1964: 271, 593).

10. Rockefeller Archives, RG 1.1, Series 200, Box 98, Folder 1189.

11. Rockefeller Archives, RG 1.1, Series 200, Box 98, Folder 1192.

12. "Mr. Charles Kettering and the Sloan-Kettering Institute," Rhoads papers, Rockefeller Archives, RG 608, Series 1, Box 4, Folder 32.

13. In the 1920s a technician who administered radium at the Memorial Hospital died from leukemia (Hayes-Martin Collection, Rockefeller Archives, RG 500, Box 6, Folder 106). Likewise, during World War I industrial workers luminized watches with radium, and they eventually fell sick from the exposure.

14. Moss (1989: 65), following Considine (1959), states that the gift was a hundred thousand dollars. Radiotherapy historian Del Regato states that the offer was for a third of a million dollars and several grams of radium (1993: 69). According to Del Regato, the offer included dedicating the hospital to the exclusive care of cancer patients, establishing an affiliation with Cornell (where Ewing had been a professor), and making Ewing "acting head as well as pathologist of the institution" (1993: 69). Original sources in the Hayes-Martin collection do not add much new information on this question (Hayes-Martin Collection, Rockefeller Archives, RG 500, Box 3, Folder 48).

15. On Douglas and his daughter, see Hayes-Martin Collection, Rockefeller Archives, RG 500, Box 3, Folder 48. Del Regato (1993: 236) reports that Douglas supported the use of radium to treat his daughter, and he asked to reserve some for his family's personal use. Moss (1989: 66) suggests that Douglas's death from anemia might have been due to radium poisoning. The U.S. government was apparently interested in the project because a large part of the ores being mined were being exported to Europe (Del Regato 1993: 69). On the radium deal, see Moss (1989: 65) and Del Regato (1993: 69). A letter from the director of the Bureau of Mines confirmed that half the radium mined would go to the Memorial Hospital, presumably via Douglas; see Hayes-Martin Collection, Rockefeller Archives, RG 500, Box 6, Folder 107, letter of June 27, 1917. On the sobriquet "radium hospital," see Rusch (1985: 393). It is likely that Ewing and Douglas purchased radium from Curie on their trip to Europe in 1913. On Curie, see Hayes-Martin Collection, Rockefeller Archives, RG 500, Box 3, Folder 48, and Box 6, Folder 101.

16. See Martin (1944) and Hayes-Martin Collection, Rockefeller Archives, RG 500, Box 6, Folder 106.

17. Studer and Chubin (1980: 21), citing Marshino (1944: 432), say that half of the $400,000 appropriation was designated for radium. According to records of a House debate, the amount may have been two hundred thousand dollars over a five-year period (Yaremchuk 1977: 97). The house debates collected by Yaremchuk suggest that the representatives' and senators' main source of information was the medical community.

18. "Industrial Science and Cancer Research," Rhoads papers, Rockefeller Archives, RG 608, Series 1, Box 4A, Folder 40C.

19. "Mr. Charles Kettering and the Sloan-Kettering Institute," Rhoads papers, Rockefeller Archives, RG 608, Series 1, Box 4, Folder 32.

20. "Frontal Attack," *Time,* July 27, 1949, pp. 66–75.

21."The Next Half Century in Cancer Research," Rhoads papers, Rockefeller Archives, RG 608, Series 1, Box 2, Folder 13.

22. Rhoads papers, Rockefeller Archives, RG 608, Series 1, Box 3, Folder 23.

23. "The Next Half Century in Cancer Research," Rhoads papers, Rockefeller Archives, RG 608, Series 1, Box 2, Folder 13.

24. Miner (1952: 872–73); "Virus Conference Report," Rhoads papers, Rockefeller Archives, RG 608, Series 1, Box 4, Folder 40.

25. "Viruses in Tumor Therapy" (paper presented at the Second National Cancer Conference, American Cancer Society, 1953), Rhoads papers, Rockefeller Archives, RG 608, Series 1, Box 2, Folders 17 and 20.

26. "Industrial Science and Cancer Research," Rhoads papers, Rockefeller Archives, RG 608, Series 1, Box 4A, Folder 40C.

27. "The Pharmaceutical Industry and Cancer Research," Rhoads papers, Rockefeller Archives, RG 608, Series 1, Box 4A, Folder 40C.

28. "Mr. Charles Kettering and the Sloan-Kettering Institute," Rhoads papers, Rockefeller Archives, RG 608, Series 1, Box 4, Folder 32.

29. Rhoads papers, Rockefeller Archives, RG 608, Series 1, Box 4A, Folder 40C, No. 30.

30. "Mr. Charles Kettering and the Sloan-Kettering Institute," Rhoads papers, Rockefeller Archives, RG 608, Series 1, Box 4, Folder 32.

31. For example, Diller was able to do her immunological research as a cytologist in the Department of Chemotherapy of the Institute for Cancer Research, and Rusch (1985: 394) mentions that Rhoads supported at least some research in virology and immunology. Viral research did not receive much attention until the 1950s and 1960s, after the successes of the polio vaccination and the growth of a body of research on animal oncoviruses (Shimkin 1977: 596).

32. Mullins (1972) and other studies reviewed in Hess (1997).

33. See also the recent work on ecologies of knowledge (Star 1995).

34. For examples of Béchamp in the alternative medicine literature today, see Bird (1990); Cantwell (1990); Cournoyer (1991); Enby (1990); and Lynes (1987). See also the biography by Nonclerq (1982).

35. Braun (1947); Klieneberger-Nobel (1949, 1951); and Dienes and Weinberger (1951).

36. Frobisher (1928). I am following Amsterdamska's selection of passages here (1991: 213). After consulting the original literature, I agree that these were two of the strongest statements against the filtrationist/cyclogenist position.

37. Wolbach (1947: 336–39) on Zinsser's research. Typhoid is caused by a member of the salmonella group, and typhus by a rickettsia; both are gram-negative bacteria, although rickettsias share characteristics with viruses.

38. Bernheim (1948) and Corner (1964: 216). Zinsser had criticized the Welch protégé Flexner; however, from other sources, Welch and Zinsser appeared to be on good terms (Flexner and Flexner 1941: 169; Fleming 1987: 133).

39. Löhnis worked at the U.S. Department of Agriculture, and Hadley had worked at the Rhode Island Agricultural Experiment Station before moving on to the University of Michigan Medical School (Amsterdamska 1991: 197, 204). However, in Michigan the agricultural university is Michigan State University, not the University of Michigan. Ralph Mellon, another cyclogenist, had a medical position but was at the nonelite Highland Hospital in Rochester, New York. The Cornell group of cyclogenists was at the Department of Public Health and Preventive Medicine at the medical school in New York City. Kendall was at Northwestern and Rosenow, a supporter of the old focal-infection theory, was at the Mayo Clinic. Thus, the agricultural dimension that Harwood suggests may exist for geneticists may not apply to the case of bacteriologists, but there may be a division along the lines of elite/nonelite background and position. Harwood stresses the exceptions to this pattern, the preliminary nature of his observation for the United States, and the need for future research. An even greater caution would apply to this tentative suggestion that I am developing from Amsterdamska's research, and more research would be necessary.

40. I have not made any attempt to answer the question of the accuracy of Mattman's perception. This would require comparing gender ratios in CWD research and mainstream bacteriology (which also had substantial participation from women) for the post–World War II years.

41. There is also a sexuality angle to the cultural meaning of these queer bacteria. This linkage is implicit in the autobiographical narrative of Alan Cantwell, a dermatologist who became a friend of Livingston, a supporter of the microbial theory of cancer, and later the advocate of a controversial theory about the origin of AIDS. His narrative openly discusses his sexuality as a gay man in the age of AIDS, and at one point he notes, "I was forty-five and I was beginning to understand my purpose in life, and why I was 'different' from the rest. I

knew I had to join Virginia in showing the cancer microbe to the world" (1990: 100).

42. For example, Livingston et al. (1950); Livingston, Alexander-Jackson, and Smith (1953); Livingston (1949, 1955); and Livingston and Alexander-Jackson (1965a, 1965b).

43. However, see the very related discussion of killer T-cells and martial imagery in Martin (1994) that has influenced my discussion here. Moss (1995: 22) also points to the "macho" nature of the military language of cancer therapy.

44. The relationship between the Calvinist religious heritage and the scientific and medical reform movement of the early twentieth century remains to be explored in detail. The background of Rockefeller and Carnegie, not to mention Rockefeller philanthropy organizer Gates and Welch, suggests that the connection might indeed be a rich one. On Welch's New England religious background see Fleming (1987) and Flexner and Flexner (1941).

45. The controversy involves his use of slave women to learn a gynecological technique and his quarrels, much later in his career, with the trustees of the Women's Hospital in New York. Compare, for example, Moss (1989: 47) with the more sympathetic portrait by McGregor (1989).

46. Associated Press report, *Albany Times-Union,* April 24, 1996, p. 5.

47. Interview, July 1995. Koch did not use hydrogen peroxide except at one point during his research, and he expressed concern over the dangers of ozone. "Instead," according to Treiger, "he thought that the way he handled free radicals was not dangerous because his products were used in minidoses, leading to the catalysis of intracellular reactions in which free radicals were formed to act physiologically, imitating nature, in order to neutralize pathogens. He recognized the necessity of neutralizing intracellular free radicals—in which he would later be supported by nutritional and orthomolecular therapies." (Personal correspondence, Dr. Jayme Treiger, January 15 and February 4, 1996).

48. Interview with William Fry and Geronimo Rubio, September 3, 1995, and visit to their clinic, September 6, 1995. Fry had studied under Rife's partner John Crane.

49. Houston (1989: 24); Office of Technology Assessment (1990: ch. 6), Moss (1989: 260–61).

NOTES TO CHAPTER 4

1. Gregory (1952) defended the viral theory; however, he used antibiotics in treatment and his descriptions of the virus included a cell wall, nucleus, and cytoplasm, so it is likely that he was observing bacteria or, if he really observed a nucleus, the prokaryotic fungi. Clark (1993) argues for a parasitic theory of cancer, but the scientific support is weak. Wyburn-Mason (1964) argued that the protozoan *Entamoeba limax* sets up precursor conditions that lead to cancer and that antiproto-

zoal drugs are effective in the treatment of cancer. Subsequent studies suggest that he had wrongly identified cell-wall deficient bacteria as protozoa (Chapdelaine 1996). Price and Bulmer (1972) found that the yeast *Cryptococcus neoformans* produced tumors when inoculated into mice, and White (1965) also supported a connection between yeast and malignancy.

2. Coley and Rife are not included because Coley did not publish much on bacterial pathogens as etiological agents of cancer, and Rife did not leave behind adequate and accessible documentation of his research. Coley's therapy is also a general bacterial therapy, not one specifically linked to a purported cancer microbe. As for the Europeans, the French and German researchers tended not to publish in peer-reviewed journals, and I also faced a problem of access.

3. Mattman (1993: 111–12) and Domingue, ed. (1982). See also Livingston and Livingston (1972).

4. See Braun (1947); Dienes and Weinberger (1951); and Klieneberger-Nobel (1949, 1951).

5. Personal correspondence, November 22, 1995. For the relevant studies, see Green, Heidger, and Domingue (1974a, 1974b); Domingue (1995, 1996); and Domingue et al. (1995).

6. For a popular review of Brown's work, see Scammell (1993). Medical citations and the importance of Brown in the bacterial etiology theory of arthritis are provided by Clark (1995). For the trials see Kloppenburg et al. (1995) and Tilley et al. (1995), and for examples of the controversy see Clark (1995); Galland (1995); McKendry (1995); and Paulus (1995).

7. See Proctor (1995: 225–35); Varmus and Weinberg (1993: 106); and Weinberg (1994: 167–70) for a review of this research.

8. Nuzum (1921, 1925); Crofton (1936); Inoue, Singer, and Hutchinson (1965); and Inoue and Singer (1970).

9. Green, Heidger, and Domingue (1974a, 1974b). See also the open cycle proposed by Domingue (1995, 1996) and the limited cyclical model of Bisset (1970).

10. Livingston and Allen (1948); Livingston (1949); and L'Esperance (1931).

11. Cohen and Strampp (1976); Acevedo et al. (1978). See also Affronti et al. (1976) and Maruo et al. (1979).

12. Acevedo et al. (1981) and Backus and Affronti (1981). The latter rules out one possible artifact and reviews the crown gall/plasmid transfer theory.

13. Acevedo, Campbell-Acevedo, and Kloos (1985). This paper shows the absence of CG production in some bacteria from cancer patients and its presence in some bacteria from noncancer patients. Backus and Affronti (1981) found CG in twelve of fourteen samples cultivated from malignant tissue, and none or very little in controls.

14. See, for example, Naughton et al. (1975); McManus, Naughton, and Martinez-Hernandez (1976); and later the Regelson editorial (1995).

15. Domingue et al. (1986: 97). The beta subunit requires at least six of a cluster of seven genes on chromosome nineteen (Krichevsky et al. 1995: 1034).

16. Grover, Woodward, and Odell (1995: 77). See also Carrell, Hammon, and Odell (1993) and Huth et al. (1994).

17. Probably the most cited of the researchers on mycoplasma and leukemia is Hayflick (Hayflick and Korpowski 1965; Hayflick 1969). The other 1960s studies on mycoplasma are cited in Gilbey and Pollard (1967), who present their largely negative results with germ-free leukemic mice.

18. Research reviewed by Mattman (1993: 312): Bunting (1914); De Negri and Mieremet (1913); Fraenkel (1912); and Fraenkel and Much (1910). Other contemporary work on Hodgkin's and bacteria includes Bloomfield (1915) and Torrey (1916). Cantwell (1981, 1990: 72–73) reviews the literature from the 1920s and 1930s.

19. Fleisher (1952); Carpenter et al. (1955); and Chang, Appleby, and Bennett (1974).

20. Cantwell (1981), Cantwell and Kelso (1984), and Alexander-Jackson (1954) and Seibert et al. (1970). A study of bovine lymphoma found evidence for bacteria in five of six specimens (McKay et al. 1967).

21. Diller, Donnelly, and Fisher (1967: 1402–3). The death from other infectious diseases suggests a possible variable infectious outcome similar to what Duran-Reynals (1950) claimed and to what Livingston proposed in her argument that cancer was related to the collagen diseases.

22. Gregory's book (1952: 19–20) summarizes his research presented in medical publications, e.g., Gregory (1948, 1949, 1950a, 1950b, 1950c, 1951). If he in fact was observing nuclei rather than some other interior body, the organism might be classified as a fungus. Another apparently independent research report was Weinman et al. (1968).

23. See Nauts (1975, 1976, 1980); Zheren and Nauts (1991); Havas, Groesbeck, and Donnelly (1958); Havas, Donnelly, and Levine (1960); Havas et al. (1993); and Johnston (1962).

24. American Cancer Society (1992), citing National Cancer Institute, Cancer Statistics Branch. The 1983–90 statistics are from Wing, Tong, and Bolden (1995: 27).

25. "Cassileth, Claiming 'Goofs,' Refuses to Resign," *Cancer Chronicles,* September 1993, p. 4. See also Falcone (1994: 81–83).

26. "Cures or 'Quackery'?" *U.S. News and World Report,* July 17, 1995, p. 49.

27. I do not intend to question the competence of the researchers. Rather, I intend to point to certain aspects of the interpretation, measurement, or design of the study that could have been improved or changed, and that would make a more convincing comparison in a future study.

28. These questions have emerged elsewhere (e.g., Lerner 1994: 330) and from the Livingston Clinic (interview with Patricia Huntley, September 1995).

29. Seibert et al. (1973) and Seibert and Davis (1977). These may be two different studies, but the design is almost identical so I am led to conclude that they may be two reports of the same study.

30. See my discussion in Hess (1995: 27–32), which reviews work by Gilbert and colleagues (Biology and Gender Study Group 1989); Hubbard (1990); Martin (1991); and others.

31. Keller (1985) and Manning (1983), extended in Hess (1995: 27–32).

NOTES TO CHAPTER 5

1. For an account of the controversy, see Proctor (1995: chs. 3–6) and Wright (1991). Sample articles in the controversy include Coggon and Inskip (1994); Davis et al. (1990); Davis, Dinse, and Hoel (1994a, 1994b); Davis and Hoel (1992); Doll and Peto (1987: 4.94–123); Epstein (1993); and Muir, Fraumeni, and Doll (1994).

2. Eisenberg et al. (1993). The pattern of use of alternative medicine may be a long-term phenomenon. For example, Beale (1939: 210) reports on a survey of YWCA members that found that 772 went to osteopaths, 120 to chiropractors, 187 to Christian Scientists, and 125 to medical doctors. Likewise, there are parallels between the Thompsonian movement of the nineteenth century and aspects of alternative medicine today.

3. On users demographics and patterns, see Cassileth et al. (1984); Furnham and Forey (1994); and Sharma (1992).

4. The phrase "medical Vietnam" as a sobriquet for the war on cancer has been attributed to Stanford University President Donald Kennedy (Proctor 1995: 4). It refers to the quagmire of research and treatment spending, and the huge toll in human lives. In a previous book (Hess 1993) I suggested that the New Age movement in the United States was linked to the baby boomer generation; it now seems clear that as the boomers age they have shown increasing interest in nutrition and alternative medicine.

5. I have been unable to document the primary source for this commonly made attribution. According to attorney Clinton Miller, who attempted to locate the primary source but failed to do so, the task of finding Rush's original statement may not be an easy one.

6. Committee for Freedom of Choice in Medicine (1995: 640); also Houston (1989: 51).

7. Another type of change is to ease restrictions on the importation of drugs and devices from other countries. For example, in response to activism from AIDS patients, in 1988 the FDA eased restrictions so that patients could import drugs from foreign countries for personal use.

8. They were able to demonstrate from available statistics in the 1980s that about a third of the grants go to these organizations, but a 1976 ACS report suggests

that the figure may be as high as 89 percent (Bennett and DiLorenzo 1994: 162–63).

9. Markle and Peterson (1980, 1987); Moss (1989: 150–51); and Peterson and Markle (1979a, 1979b).

10. See Pauling (1993); Richards (1981) for a more historical account, and Riordan et al. (1994) for subsequent research.

11. Kamen (1993) suggested that the clinical trials may not have excluded the incompatibles of alcohol, sleeping pills, and tranquilizers, and that patients may have received prior chemotherapy in conflict with the protocol. The clinical studies were published in the June, 1994, issue of *Journal of Clinical Oncology*. For a subsequent response to the charge that incompatibles were not excluded, see Kosty et al. (1995).

12. The suppression of Burzynski's work is chronicled in the newsletter *Options: Revolutionary Ideas in the War on Cancer* (People Against Cancer, P.O. Box 10, Otho, Iowa 50569). For the purposes here the most relevant article is the one where Moss is quoted: "Burzynski Charges National Cancer Institute with Misconduct: NCI Sponsored Trials Halted!" (October, 1995, p. 4).

13. For example, I am thinking of the apparently rapidly changing consensus in cancer treatment that is increasingly opening the door to adjuvant nutritional therapies, as in the national symposium sponsored by the American College of Nutrition and the Cancer Treatment Research Foundation in September, 1995.

14. Western psychotherapies occupy an ambiguous middle ground. There is a consensus that patients with a better attitude and social support generally do better, so at this level they are not really alternative therapies, but standard adjuvant therapies. However, other claims, such as the occasional claims that hypnosis or suggestion can relieve chronic, physical diseases, are complicated and beyond the scope of the general issue that I am discussing here.

Bibliography

Acevedo, Hernan, Elizabeth Campbell-Acevedo, and Wesley Kloos

1985 "Expression of Human Choriogonadotropin-like Material in Coagulase-Negative *Staphylococcus* Species." *Infection and Immunity* 50(3): 860–68.

Acevedo, Hernan, Samuel Koide, Malcolm Slifkin, Takeshi Maruo, and Elizabeth Campbell-Acevedo

1981 "Choriogonadotropin-like Antigen in a Strain of *Streptococcus faecalis* and a Strain of *Staphylococcus simulans:* Detection, Identification, and Characterization." *Infection and Immunity* 31(1): 487–94.

Acevedo, Hernan, Alexander Krichevsky, Elizabeth Campbell-Acevedo, Joyce Galyon, Mary Jo Buffo, and Robert Hartsock

1995a "Flow Cytometry Method for the Analysis of Membrane-Associated Human Chorionic Gonadotropin, Its Subunits, and Fragments on Human Cancer Cells." *Cancer* 69: 1818–28.

1995b "Expression of Membrane-Associated Human Chorionic Gonadotropin, Its Subunits, and Fragments by Cultured Human Cancer Cells." *Cancer* 69: 1829–42.

Acevedo, Hernan, Matias Pardo, Elizabeth Campbell-Acevedo, and Gerald Domingue

1987 "Human Choriogonadotropin-like Material in Bacteria of Different Species: Electron Microscopy and Immunocytochemical Studies with Monoclonal and Polyclonal Antibodies." *Journal of General Microbiology* 133: 783–91.

Acevedo, Hernan, Malcolm Slifkin, Gail Pouchet, and Matias Pardo

1978 "Immunohistochemical Localization of a Chorionic Gonadotropin-like Protein in Bacteria Isolated from Cancer Patients." *Cancer* 41: 1217–29.

Acevedo, Hernan, Malcolm Slifkin, Gail Pouchet-Melvin, and Elizabeth Campbell-Acevedo

1979 "Choriogonadotropin-like Antigen in an Anaerobic Bacterium, *Eubacterium Lentum,* Isolated from a Rectal Tumor." *Infection and Immunity* 24 (3): 920–24.

Acevedo, Hernan, Jennifer Tong, and Robert Hartsock

1995 "Human Chorionic Gonadotropin-Beta Subunit Gene Expression in Cultured Human Fetal and Cancer Cells of Different Types and Origins." *Cancer* 76: 1467–75.

Adamson, C. A.

1949 "Bacteriological Study of Lymph Nodes: Analysis of Postmortem Specimens with Particular Reference to Clinical, Serological, and Histopathological Findings." *Acta Medica Scandinavica. Supplementum* 227: 1–21.

Affronti, Lewis, Linda Grow, and Fred Begell

1975 "Characterization of Bacterial Tumor Isolates." *Proceedings of the Federation of American Societies for Experimental Biology* (March 1), p. 1043.

Affronti, Lewis, Linda Grow, R. Brumbaugh, and K. Orton

1976 "Abstract." *Abstracts of the Annual Meeting of the American Society for Microbiology* E56, p. 72.

Alexander-Jackson, Eleanor

1954 "A Specific Type of Microorganism Isolated from Animal and Human Cancer: Bacteriology of the Organism." *Growth* 18: 37–51.

1966 "Mycoplasma (PPLO) Isolated from Rous Sarcoma Virus." *Growth* 30: 199–228.

1970 "Ultraviolet Spectrogramic Microscope Studies of Rous Sarcoma." *Annals of the New York Academy of Sciences* Vol. 174, Art. 2, p. 765.

American Cancer Society (ACS)

1966 *Unproven Methods of Cancer Treatment.* Atlanta: American Cancer Society.

1968 "Unproven Methods of Cancer Management: The Livingston Vaccine." *CA—A Cancer Journal for Clinicians* 18: 46–47.

1990 "Unproven Methods of Cancer Management: Livingston-Wheeler Therapy." *CA—A Cancer Journal for Clinicians* 40: 103–8.

1992 *Cancer Facts and Figures.* Atlanta: American Cancer Society.

Ames, Bruce

1995 "The Causes and Prevention of Cancer." *Proceedings of the National Academy of Sciences of the United States of America* 92 (June 6): 5258–65.

Amsterdamska, Olga

1987 "Medical and Biological Constraints: Early Research on Variation in Bacteriology." *Social Studies of Science* 17: 657–87.

1991 "Stabilizing Instability: The Controversy over Cyclogenic Theories of Bacterial Variation during the Interwar Period." *Journal of the History of Biology* 24(2): 191–222.

Appadurai, Arjun

1990 "Disjuncture and Difference in the Global Political Economy." *Public Culture* 2(2): 1–24.

Atlas, Robert

1988 *Microbiology: Fundamentals and Applications.* 2nd edition. New York: Macmillan.

Backus, Beverly, and Lewis Affronti

1981 "Tumor-Associated Bacteria Capable of Producing a Human Choriogonadotropin-like Substance." *Infection and Immunity* 32(3): 1211–15.

Baer, Hans

1987 "Divergence and Convergence in Two Systems of Manual Medicine: Osteopathy and Chiropractic in the United States." *Medical Anthropology Quarterly* 1(2): 176–93.

1989 "The American Dominative Medical System as a Reflection of Social Relations in the Larger Society." *Social Science and Medicine* 28(11): 1103–12.

Bailey, Herbert

1958 *A Matter of Life and Death: The Incredible Story of Krebiozen.* New York: G. P. Putnam's Sons.

Barbosa, Lívia

1995 "The Brazilian *Jeitinho.*" In David Hess and Roberto DaMatta (eds.), *The Brazilian Puzzle: Culture on the Borderlands of the Western World.* New York: Columbia University Press.

Barnes, Barry

1977 *Interests and the Growth of Knowledge.* London: Routledge.

1981 "On the 'Hows' and 'Whys' of Cultural Change." *Social Studies of Science* 11: 481–98.

Barnes, Barry, and Donald MacKenzie

1979 "On the Role of Interests in Scientific Change." In Roy Wallis (ed.), *On the Margins of Science. Sociological Review Monograph No. 27.* Keele, Staffordshire: University of Keele.

Barnes, Barry, and Steven Shapin (eds.)

1979 *Natural Order.* Beverly Hills: Sage.

Bastide, Roger

1978 *The African Religions of Brazil.* Baltimore: Johns Hopkins University Press.

Beale, Morris

1939 *Medical Mussolini.* Washington, D.C.: Columbia Publishing Co.

Beard, John

1911 *The Enzyme Treatment of Cancer.* London: Chatto and Windus.

Béchamp, Antoine

1911 *The Blood and Its Third Anatomical Element.* Philadelphia: Boericke and Tafel.

Beinhauer, Lawrence, and Ralph Mellon

1938 "Pathogenesis of Noncaseating Epithelioid Tuberculosis of Hypoderm and Lymph Glands." *Archives of Dermatology and Syphillus* 37: 451–60.

Bennett, James, and Thomas DiLorenzo

1994 *Unhealthy Charities.* New York: Basic Books.

Berg, Richard, and William Coley
1932 "Experimental Production of Several Varieties of Bone Sarcoma by Intramedullary Injections of the Virus of the Filterable Fowl Endothelioma Tumor." *American Journal of Surgery* 15: 441–61.

Berliner, Howard
1985 *A System of Scientific Medicine.* New York and London: Tavistock.

Bernheim, Bertram
1948 *The Story of Johns Hopkins.* New York: McGraw-Hill.

Biology and Gender Study Group
1989 "The Importance of Feminist Critique for Contemporary Cell Biology." In Nancy Tuana (ed.), *Feminism and Science.* Bloomington, Ind.: Indiana University Press.

Bird, Christopher
1990 *The Persecution and Trial of Gaston Naessens.* Tiburon, Calif.: H. J. Kramer.

Bisset, Kenneth A.
1969 "Bacteriology's Next Battlefield." *New Scientist* 42(653): 580–81.
1970 *The Cytology and Life-History of Bacteria.* Edinburgh: E. & S. Livingstone.

Bloomfield, A. L.
1915 "The Bacterial Flora of Lymphatic Glands." *Archives of Internal Medicine* 16: 197–204.

Bloor, David
1991 *Knowledge and Social Imagery.* 2nd edition. Chicago: University of Chicago Press.

Blumenthal, Andrea Kathryn
1991 *Leadership in a Medical Philanthropy: Simon Flexner and the Rockefeller Institute for Medical Research.* Ph.D. dissertation, Drew University, Madison, N.J.

Boesch, Mark
1960 *The Long Search for the Truth about Cancer.* New York: G. P. Putnam's.

Borkin, Joseph
1978 *The Crime and Punishment of I. G. Farben.* New York: Free Press.

Bowker, Geof, and Bruno Latour
1987 "A Booming Discipline Short of Discipline." *Social Studies of Science* 17: 715–48.

Braun, Werner
1947 "Bacterial Dissociation." *Bacteriological Review* 10: 75–114.

Brown, E. Richard
1979 *The Rockefeller Medicine Men: Medicine and Capitalism in America.* Berkeley and Los Angeles: University of California Press.

Brown, M. L.

1990 "Special Report: The National Economic Burden of Cancer: An Update." *Journal of the National Cancer Institute* 82: 1811–14.

Bud, R. F.

1978 "Strategy in American Cancer Research After World War II: A Case Study." *Social Studies of Science* 8: 429–59.

Bunting, C. H.

1914 "The Blood-Picture in Hodgkin's Disease, second paper." *Bulletin of Johns Hopkins Hospital* 25: 173–84.

Callon, Michel

1986 "Some Elements of a Sociology of Translation: Domestication of the Scallops and Fishermen." In John Law, (ed.), *Power, Action, and Belief.* Sociological Review Monograph No. 32 (University of Keele). London: Routledge.

Callon, Michel, and John Law

1982 "On Interests and their Transformation: Enrollment and Counterenrollment." *Social Studies of Science* 12: 615–25.

Cantwell, Alan, Jr.

1981 "Histologic Observations of Variably Acid Fast Coccoid Forms Suggestive of CWD Bacteria in Hodgkin's Disease, 4 Cases." *Growth* 45: 168–87.

1990 *The Cancer Microbe: The Hidden Killer in Cancer, AIDS, and Other Diseases.* Los Angeles: Aries Rising Press.

Cantwell, Alan, Jr., and D. W. Kelso

1971 "Acid-Fast Bacteria in Scleroderma and Morphea." *Archives of Dermatology* 104: 21–25.

1984 "Variably Acid-Fast Bacteria in a Fatal Case of Hodgkin's Disease." *Archives of Dermatology* 120: 401–2.

Carnap, Rudolf

1995 *An Introduction to the Philosophy of Science.* Mineola, N.Y.: Dover Books. (Reprint of *Philosophical Foundations of Physics.* New York: Basic Books, 1966.)

Carpenter, C. M., E. L. Nelson, E. L. Lehman, D. H. Howard, and G. Primbs

1955 "The Isolation of Unidentified Pleomorphic Bacteria from the Blood of Patients with Chronic Illness." *Journal of Chronic Disease* 2: 156–61.

Carrell, Douglas, M. Elizabeth Hammon, and William Odell

1993 "Evidence for an Autocrine/Paracrine Function of Chorionic Gonadotropin in *Xanthomonas maltophilia.*" *Endocrinology* 132(3): 1085–89.

Carter, James P.

1993 *Racketeering in Medicine.* Norfolk, Va.: Hampton Roads.

Cassileth, Barrie, Edward Lusk, DuPont Guerry, Alicia Blake, William Walsh, Lauren Kascius, and Delray Schultz

1991 "Survival and Quality of Life Among Patients Receiving Unproven as Compared with Conventional Cancer Therapy." *The New England Journal of Medicine* April 25, pp. 1180–85.

Cassileth, Barrie, Edward Lusk, Thomas Strouse, and Brenda Brodenheimer

1984 "Contemporary Unorthodox Treatments in Cancer Medicine." *Annals of Internal Medicine* 101(1): 105–12.

Chang, J. C., J. Appleby, and J. M. Bennett

1974 "Nitroblue Tetrazolium Test in Hodgkin's Disease and Other Malignant Lymphomas." *Archives of Internal Medicine* 133: 401–3.

Chapdelaine, Perry

1996 "History of the Roger Wyburn-Mason and Jack M. Blount Foundation for Eradication of Rheumatoid Disease." *Townsend Letter for Doctors and Patients* January, pp. 76–81.

Clark, George

1953 "Successful Culturing of Glover's Cancer Organism and Development of Metastasizing Tumors in Animals Produced by Cultures from Human Malignancy." *Atti del VI Congresso Internazionale di Microbiologia* Vol. 6, sec. 17A: 41–49.

Clark, Harold

1995 "Letter to the Editor." *Annals of Internal Medicine* 123(5): 393.

Clark, Hulda

1993 *The Cure for All Cancers.* West Haven, Conn.: Twin Press.

Coggon, David, and Hazel Inskip

1994 "Is There an Epidemic of Cancer?" *British Medical Journal* March 308: 705–8.

Cohen, Herman, and Alice Strampp

1976 "Bacterial Synthesis of a Substance Similar to Human Chorionic Gonadotropin." *Proceedings of the Society for Experimental Biology and Medicine* 152: 408–10.

Cole, Stephen

1992 *Making Science.* Cambridge: Harvard University Press.

Coley, William

1925 "Some Clinical Evidence in Favor of the Extrinsic Origin of Cancer." *Surgery, Gynecology, and Obstetrics* 40: 353–59.

1926 "The Cancer Symposium at Lake Mohonk." *American Journal of Surgery* (New Series) 1: 222–25.

1928 "Some Observations on the Problem of Cancer Control." *American Journal of Surgery* (New Series) 4: 663–82.

1931 "Some Thoughts on the Problem of Cancer Control." *American Journal of Surgery* (New Series) 14: 605–19.

Collins, Harry

1985 *Changing Order: Replication and Induction in Scientific Practice.* Beverly Hills: Sage.

Committee for Freedom of Choice in Medicine
1995 "Plan for Health Care Reform." In Michael Culbert (ed.), *Medical Armageddon.* Vols. 3 & 4. San Diego: C&C Communications.

Considine, Robert
1959 *That Many May Live.* New York: Memorial Center for Cancer and Allied Diseases.

Corner, George
1964 *A History of the Rockefeller Institute, 1901–1953: Origins and Growth.* New York: Rockefeller Institute Press.

Coulter, Harris
1987 *Divided Legacy: The Conflict Between Homeopathy and the American Medical Association. Vol. 3: Science and Ethics in America Medicine, 1800–1914.* 2nd edition. Richmond, Calif.: North Atlantic Books.

Council on Pharmacy and Chemistry
1951 "Report of the Council." *Journal of the American Medical Association* 147(9): 864–73.

Cournoyer, Cynthia
1991 *What About Immunizations? Exposing the Vaccine Philosophy.* Santa Cruz, Calif.: Nelson's Books.

Cowley, Geoffrey
1995 "Going Mainstream." *Newsweek* June 16, pp. 56–57.

Crofton, William M.
1936 *The True Nature of Viruses.* London: Staples Press.

Culler, Jonathan
1982 *On Deconstruction.* Ithaca: Cornell University Press.

Culliton, Barbara
1974 "Virus Cancer Program: Review Panel Stands by Criticism." *Science* 184(Apr. 12): 143–45.

DaMatta, Roberto
1991 *Carnivals, Rogues, and Heroes.* Notre Dame: University of Notre Dame Press.

Davis, Devra Lee, Gregg Dinse, and David Hoel
1994a "Decreasing Cardiovascular Disease and Increasing Cancer among whites in the United States from 1973 through 1987." *Journal of the American Medical Association* 271(6): 431–37.
1994b "In Reply." *Journal of the American Medical Association* 272(3): 199–200.

Davis, Devra, and David Hoel
1992 "Figuring Out Cancer." *International Journal of Health Services* 22(3): 447–53.

Davis, Devra, David Hoel, John Fox, and Alan Lopez
1990 "International Trends in Cancer Mortality in France, West Germany, Italy, Japan, England, and Wales, and the USA." *Lancet* August 25: 474–81.

Del Regato, Juan

1993 *Radiation Oncologists: The Unfolding of Radiology.* Reston, Va.: Radiology Centennial.

De Negri, E., and C. W. G. Mieremet

1913 "Zür Aetiologie des Malignen Granuloms." *Zentralblat für Bakteriologie, Parasitenkunde, Infektionskrankheiten, und Hygiene. Abteilung 1 Originale* 68: 292–308.

De Tocqueville, Alexis

1969 *Democracy in America.* Garden City, N.Y.: Doubleday.

Dienes, Louis, and Howard Weinberger

1951 "The L Forms of Bacteria." *Bacteriological Review* 15: 245–88.

Diller, Irene

1962a "Growth and Morphological Variability of Three Similar Strains of Intermittently Acid-Fast Organisms Isolated from Mouse and Human Malignant Tissues." *Growth* 26(3): 181–208.

1962b "Three Similar Strains of Pleomorphic Acid-Fast Organisms Isolated from Rat and Mouse Tissues and from Human Blood." *American Review of Respiratory Disease* 86(6): 932–35.

1974 "Tumor Incidence in ICR/Albino and C57/N16JNIcr Male Mice Injected with Organisms Cultured from Mouse Malignant Tissues." *Growth* 38: 507–17.

Diller, Irene, and Andrew Donnelly

1970 "Experiments with Mammalian Tumor Isolates." *Annals of the New York Academy of Sciences* 174(2): 655–74.

Diller, Irene, Andrew Donnelly, and Mary Fisher

1967 "Isolation of Pleomorphic, Acid-fast Organisms from Several Strains of Mice." *Cancer Research* 27 (Part 1): 1402–8.

Diller, Irene, and G. Medes

1964 "Isolation of a Pleomorphic Acid-Fast Organism from Liver and Blood of Carcinogen-Fed Rats." *American Review of Respiratory Diseases* 90(1): 126–28.

Doll, R., and R. Peto

1987 "Epidemiology of Cancer." In D. J. Weatherall, J. G. G. Ledingham, and D. A. Warrel (eds.), *Oxford Textbook of Medicine.* New York: Oxford University Press.

Domingue, Gerald

1982 "Filterable, Cell-Associated Cell Wall-Deficient Bacteria in Renal Diseases." In Gerald Domingue (ed.), *Cell Wall-Deficient Bacteria: Basic Principles and Clinical Significance.* Reading, Mass.: Addison-Wesley.

1995 "Electron Dense Cytoplasmic Particles and Chronic Infection: A Bacterial Pleomorphy Hypothesis." *Endocytobiosis and Cell Research* 11: 19–40.

1996 "Pleomorphic Cell Wall-Defective Bacteria as Cryptic Agents of Disease." In *Pleomorphism in Biology and Medicine.* Philadelphia, Penn.: Center for Frontier Sciences.

Domingue, Gerald (ed.)

1982 *Cell Wall Deficient Bacteria.* Reading, Mass.: Addison-Wesley.

Domingue, Gerald, Hernan Acevedo, John Powell, and Vernon Stevens

1986 "Antibodies to Bacterial Vaccines Demonstrating Specificity for Human Choriogonadotropin (hCG) and Immunochemical Detection of hCGlike Factor in Subcellular Bacterial Fractions." *Infection and Immunity* 53(1): 95–98.

Domingue, Gerald, Gamal Ghoniem, Kenneth Bost, Cesar Fermin, and Liset Human

1995 "Dormant Microbes in Interstitial Cystitis." *Journal of Urology* 153: 1321–26.

Downey, Gary, Joe Dumit, and Sharon Traweek (eds.)

1997 *Cyborgs and Citadels.* Santa Fe: School for American Research.

Drinkard, John

1996 "Study: Drug, Tobacco Firms Aid Effort to Curb FDA." Associated Press report, July 24, *Schenectady Gazette.*

Dubos, René

1950 *Louis Pasteur: Free Lance of Science.* Boston: Little, Brown, and Co.

Duhem, Pierre

1982 *The Aim and Structure of Physical Theory.* Princeton: Princeton University Press.

Duran-Reynals, Francisco

1933 "Reaction of Transplantable and Spontaneous Tumors to Blood Carried Bacterial Toxins in Animals Unsusceptible to the Shwartzman Phenomenon." *Proceedings of the Society for Experimental Biology and Medicine* 31: 341–44.

1950 "Neoplastic Infection and Cancer." *American Journal of Medicine* 8(4): 440–511.

Eisenberg, David, Ronald Kessler, Cindy Foster, Frances Norlock, David Calkins, and Thomas Delbanco

1993 "Unconventional Medicine in the United States." *New England Journal of Medicine* 328(4): 246–52.

Ellerman, V., and O. Bang

1908 "Experimentelle Leukämie bei Hühnern." *Zentralblat für Bakteriologie, Parasitenkunde, Infektionskrankheiten, und Hygiene. Abteilung 1 Originale* 46: 595–609.

Enby, Erik, with Michael Sheehan

1990 *Hidden Killers: The Revolutionary Medical Discoveries of Gunther Enderlein.* Saratoga, Calif.: S&G Communications.

Enderlein, Günther

1925 *Bakterien-Cyclogenie.* Berlin and Leipzig: Walter de Gruyter and Co.

Epstein, Samuel

1993 "Evaluation of the National Cancer Program and Proposed Reforms." *International Journal of Health Services* 23(1): 15–44.

Epstein, Samuel, Eula Bingham, David Rall, and Irwin Bross

1992 "Losing the 'War Against Cancer': A Need for Public Policy Reforms." *International Journal of Health Services* 22(3): 455–69.

Eurogast Study Group

1993 "An International Association Between *Helicobacter Pylori* Infection and Gastric Cancer." *Lancet* 341: 1359–62.

Ewing, James

1919 *Neoplastic Diseases.* 2nd edition. Philadelphia: W. B. Saunders and Co.

Falcone, Ron

1994 *The Complete Guide to Alternative Cancer Therapies.* New York: Carol Communications/Citadel Press.

Fleisher, M. S.

1952 "Significance of Diptheroid Microorganisms in Blood Cultures from Human Beings." *American Journal of Medical Science* 224: 548–53.

Fleming, Donald

1987 *William H. Welch and the Rise of Modern Medicine.* Baltimore: Johns Hopkins University Press.

Flexner, Abraham

1910 *Medical Education in the United States and Canada.* New York: Carnegie Foundation for the Advancement of Teaching.

Flexner, James

1987 *An American Saga: The Story of Helen Thomas and James Flexner.* New York: Simon and Schuster.

Flexner, Simon, and James Thomas Flexner

1941 *William Henry Welch and the Heroic Age of American Medicine.* New York: Viking Press.

Fonti, Clara

1958 *Etiopatogenese del Cancro.* Milan: Instituto Editorial Cisalpino.

Force, E. E., and R. C. Stewart

1964 "Effect of 5 Iodo-2-Deoxyuridine on Multiplication of the Rous Sarcoma Virus *in vitro.*" *Proceedings of the Society for Experimental Biology and Medicine* 116: 803–6.

Fraenkel, E.

1912 "Über die sogen. Hodgkinische Krankheir (Lymphomatosis Granulomatosa)." *Deutsche Medzinische Wochenschrift* 637–42.

Fraenkel, E., and H. Much

1910 "Über die Hodgkinische Krankheit (Lymphomatosis granulomatosa), insbesondire deren Ätiologie." *Zentralblatt für Hygiene und Umweltmedizin* 67: 159–200.

Freudenthal, Gad

1990 "Science studies in France: A Sociological View." *Social Studies of Science* 20: 353–69.

Friedmann, Friedrich

1913 *Dr. Friedmann's New Treatment for Tuberculosis.* Washington, D.C.: U.S. Printing Office.

Frobisher, Martin

1928 "On the Action of Bacteriophage in Producing Filtrable Forms and Mutations of Bacteria." *Journal of Infectious Disease* 42: 462.

Fujimura, Joan

1995 "Ecologies of Action: Recombining Genes, Molecularizing Cancer, and Transforming Biology." In Susan Leigh Star (ed.), *Ecologies of Knowledge: Work and Politics in Science and Technology.* Albany: State University of New York Press.

Fujinami, A., and K. Inamoto

1914 "Über Geschwülste bei japnischen Haushühnern, insebesondere über einen transplantablen Tumor." *Zentralblatt für Krebsforschung* 14: 94–119.

Fuller, Steve

1988 *Social Epistemology.* Bloomington, Ind.: University of Indiana Press.

1993 *Philosophy of Science and Its Discontents.* 2nd edition. New York: Guilford Press.

Furnham, A., and J. Forey

1994 "The Attitudes, Behaviors, and Beliefs of Patients of Conventional versus Complementary (Alternative) Medicine." *Journal of Clinical Psychology* 50: 458–69.

Galland, Leo

1995 "Letter to the Editor." *Annals of Internal Medicine* 123(5): 392–93.

Gerlach, Franz

1948 *Krebs und Obligäter Pilzparasitismus.* Vienna: Urban U. Schwarzenberg.

Giere, Ron

1993 "Science and Technology Studies: Prospects for an Enlightened Postmodern Synthesis." *Science, Technology, and Human Values* 18(1): 102–12.

1995 "Viewing Science." Presidential address, Philosophy of Science Association.

Gieryn, Thomas

1983a "Boundary-Work and the Demarcation of Science from Non-Science." *American Sociological Review* 48: 781–95.

1983b "Making the Demarcation of Science a Sociological Problem: Boundary-Work by John Tyndall, Victorian Scientist." *Working Papers in Science and Technology: The Demarcation between Science and Pseudoscience.* Blacksburg, Va.: Center for the Study of Science in Society, Virginia Tech.

Gilbey, Jack, and Morris Pollard
1967 "Search for Mycoplasma in Germ-free Leukemic Mice." *Journal of the National Cancer Institute* 38(2): 113–16.

Gill, Parkash, Yanto Lunardi-Iskandor, Stan Louie, Anile Tulpule, Tong Zheng, Byron Espina, Jacques Besnier, Philippe Hermans, Alexandra Levine, Joseph Bryant, and Robert Gallo
1996 "The Effect of Preparations of Human Chorionic Gonadotropin on AIDS-Related Kaposi's Sarcoma." *New England Journal of Medicine* 335(17): 1216–69.

Glover, Thomas
1926 "Progress in Cancer Research." *Canada Lancet and Practitioner* 67(5): 161–216.
1930 "The Bacteriology of Cancer." *Canada Lancet and Practitioner* 74(3): 92–111.

Glover, Thomas, and J. L. Engle
1933 "Production of a Malignant Growth in a Guinea Pig." *Public Health Reports* 48(13): 319–21.
1938 *Studies in Malignancy.* New York: Murdock Foundation.

Glover, Thomas, Michael Scott, Julian Loudon, and J. M. McCormack
1926 "A Study of the Rous Chicken Sarcoma No. 1." *Canada Lancet and Practitioner* 66(2): 49–62.

Glover, Thomas, and J. E. White
1940 *The Treatment of Cancer in Man.* New York: Murdock Foundation.

Graille, Jean-Michel
1984 *Dossier Priore: Une nouvelle affaire Pasteur?* Paris: Ed. Denoel.

Green, M. T., P. M. Heidger, Jr., and G. J. Domingue
1974a "Demonstration of the Phenomena of Microbial Persistence and Reversion with Bacterial L-Forms." *Infection and Immunology* 10: 889–914.
1974b "Proposed Reproductive Cycle for a Relatively Stable L-Phase Variant of *Streptococcus Faecalis.*" *Infection and Immunology* 10: 915–27.

Gregory, John E.
1948 "Electron Microscopic Findings in Malignant Tissue." *Experimental Medicine and Surgery* 6: 390–504.
1949 "A Review of Cancer Research." *Experimental Medicine and Surgery* 7: 289–98.
1950a "Virus as a Cause of Human and Malignant Malignancies." *Southern Medical Journal* 43 (Feb.): 124–28.
1950b "B. Subtilis as an Antibiotic in the Treatment of Cancer." *Southern Medical Journal* 43 (May): 397–403.
1950c "Virus as the Cause of Human Cancer." *Cinquième Congrès International contre le Cancer.* Paris. July. Act Vol. 8, No. 1. November.
1951 "Virus as the Cause of Human Cancer." *Michigan State Medical Journal,* April.
1952 *Pathogenesis of Cancer.* Pasadena: Freemont Foundation.

Gross, Ludwik
1983 *Oncogenic Viruses.* New York: Pergamon Press.

Grover, Sanjeev, Scott Woodward, and William Odell

1995 "Complete Sequence of the Gene Encoding a Chorionic Gonadotropin-like Protein from *Xanthomonas maltophilia*." *Gene* 156: 75–78.

Gruner, O. Cameron

1942 *Study of Blood in Cancer.* Montreal: Renouf.

Gye, William

1925a "The Aetiology of Malignant New Growths." *Lancet* July 18: 109–17.

1925b "Discussion on Filter-Passing Viruses and Cancer." *British Medication Journal* Aug. 1: 189–95.

Hadley, Phillip

1927 "The Instability of Bacterial Species with Special Reference to Active Dissociations and Transmissions." *Journal of Infectious Diseases* 40: 1–312.

1937 "Further Advances in the Study of Microbic Dissociation." *Journal of Infectious Diseases* 60: 129–62.

Haraway, Donna

1989 *Primate Visions.* London and New York: Routledge.

1991 *Simians, Cyborgs, and Women.* London and New York: Routledge.

Harding, Sandra

1986 *The Science Question in Feminism.* Ithaca, N.Y.: Cornell University Press.

1992 "After the Neutrality Ideal: Science, Politics, and 'Strong Objectivity.' " *Social Research* 59(3): 567–87.

Harvey, David

1989 *The Condition of Postmodernity.* Oxford: Blackwell.

Harwood, Jonathan

1993 *Styles of Scientific Thought.* Chicago: University of Chicago Press.

Hasumi, Kiichiro

1980 *Cancer Has Been Conquered: The Hasumi Cancer Virus Vaccines.* Tokyo: Maruzen Co., Ltd.

Haught, S. J.

1991 *Censured for Curing Cancer: The American Experience of Dr. Max Gerson.* Barrytown, N.Y.: Station Hill Press (P.U.L.S.E.).

Havas, H. Francis, Rita Axelrod, Mary Burns, Donna Murasko, and Michael Goonewardene

1993 "Clinical Results and Immunological Effects of a Mixed Bacterial Vaccine in Cancer Patients." *Medical Oncology and Tumour Pharmacotherapy* 10(4): 145–58.

Havas, H. F., A. J. Donnelly, and S. I. Levine

1960 "Mixed Bacterial Toxins in the Treatment of Tumors. III. Effect of Tumor Removal on the Toxicity and Mortality Rates in Mice." *Cancer Research* 20: 393–96.

Havas, H. F., M. E. Groesbeck, and A. J. Donnelly

1958 "Mixed Bacterial Toxins in the Treatment of Tumors. I. Methods of Preparation and Effects on Normal or Sarcoma 37-bearing Mice." *Cancer Research* 18: 141–48.

Hayflick, L.

1969 "Mycoplasmas from Malignant Tissue." In L. Hayflick (ed.), *The Mycoplasmatales and the L-Phase of Bacteria*. New York: Appleton-Century-Crofts.

Hayflick, L., and H. Korpowski

1965 "Direct Agar Isolation of Mycoplasmas from Human Leukemic Bone Marrow." *Nature* 205: 712–14.

Hess, David

1991 *Spirits and Scientists.* University Park, Penn.: Pennsylvania State University Press.

1993 *Science in the New Age.* Madison: University of Wisconsin Press.

1995 *Science and Technology in a Multicultural World*. New York: Columbia University Press.

1996 "If You're Thinking of Living in STS . . . A Guide for the Perplexed." In Gary Downey, Joe Dumit, and Sharon Traweek (eds.), *Cyborgs and Citadels.* Santa Fe: School for American Research Press.

1997 *Science Studies: An Advanced Introduction*. New York: New York University Press.

Hildenbrand, Gar, L. Christeene Hildenbrand, Karen Bradford, and Shirley Cavin

1995 "Five-Year Survival Rates of Melanoma Patients Treated by Diet Therapy after the Manner of Gerson: A Retrospective Review." *Alternative Therapies* 1(4): 29–37.

Hildenbrand, Gar, L. Christeene Hildenbrand, Karen Bradford, Dan E. Rogers, Charlotte Gerson Strauss, and Shirley Cavin

1996 "The Role of Follow-up and Retrospective Data Analysis in Alternative Cancer Management: The Gerson Experience." *Journal of Naturopathic Medicine* 6(1): 49–56.

Houston, Robert

1989 *Repression and Reform in the Evaluation of Alternative Cancer Therapies.* Washington, D.C.: Project Cure, Inc.

Hubbard, Ruth

1990 *The Politics of Women's Biology.* New Brunswick, N.J.: Rutgers University Press.

Hughes, R. A.

1994 "Focal Infection Revisited." *British Journal of Rheumatology* 33: 370–77.

Hume, E. Douglas

1932 *Béchamp or Pasteur? A Lost Chapter in the History of Biology.* London: C. W. Daniel, Co.

Huth, Jeffrey, Sheila Norton, Oksana Lockridge, Toshihiko Shikone, Aaron Hsueh, and Raymond Ruddon

1994 "Bacterial Expression and *in Vitro* Folding of the Beta-Subunit of Human Chorionic Gonadotropin and Functional Assembly of Recombinant hCG-Beta with hCG-Alpha." *Endocrinology* 135(3): 911–18.

Inoue, Sakae, and Marcus Singer

1970 "Experiments on a Spontaneously Originated Visceral Tumor in the New, *Triturus Pyrrhogaster.*" *Annals of the New York Academy of Sciences* 174(2): 729–64.

Inoue, Sakea, Marcus Singer, and Joanne Hutchinson

1965 "Causative Agent of a Spontaneously Originating Visceral Tumor in the Newt, *Triturus.*" *Nature* 205: 408–9.

Issels, Josef

1975 *Cancer: A Second Opinion.* London: Hodder and Stoughton.

Ivy, Andrew

1956 *Observations on Krebiozen in the Management of Cancer.* Chicago: H. Regnery.

Jaffe, Richard

1995 "FDA Abuse and Retaliation: Dr. Stanislaw Burzynski." *Townsend Letter for Doctors and Patients* August/September, pp. 13–16.

Janeway, Henry

1917 *Radium Therapy in Cancer at the Memorial Hospital in New York. First Report: 1915–1916.* New York: Paul Hoeber.

Jeljaszewicz, J., G. Pulverer, and W. Roszkowski

1982 *Bacteria and Cancer.* New York: Academic Press.

Johnston, Barbara

1962 "Clinical Effects of Coley's Toxin. I. A Controlled Study. 2. A Seven-Year Study." *Cancer Chemotherapy Reports* 21: 19–68.

Kamen, Jeff

1993 "Hope, Heartbreak, and Horror." *Omni* September, supplement.

Kassel, Robert, and Antonio Rottino

1955 "Significance of Diptheroids in Malignant Disease Studied by Germ-Free Techniques." *Archives of Internal Medicine* 96: 804–8.

Kellen, John, Arnost Kolin, and Hernan Acevedo

1982 "Effects of Antibodies to Choriogonadotropin in Malignant Growth, I. Rat. 3230 AC Mammary Adenocarcinoma." *Cancer* 49(11): 2300–4.

Kellen, John, Arnost Kolin, Apkar Mirakian, and Hernan Acevedo

1982 "Effects of Antibodies to Choriogonadotropin in Malignant Growth, II. Solid Transplantable Rat Tumors." *Cancer Immunology and Immunotherapy* 13: 2–4.

Keller, Evelyn Fox

1985 *Reflections on Gender and Science.* New Haven: Yale University Press.

Kempin, S., C. Cirrincione, J. Myers, B. Lee III, D. Straus, B. Koziner, Z. Arlin, T. Gee, R. Mertelsmann, C. Pinsky, E. Comacho, L. Nisce, L. Old, B. Clarkson, and H. Oettgen

1983 "Combined Modality Therapy of Advanced Lymphomas (NL): The Role of Non-Specific Immunotherapy (MBV) as Important Determinant of Response and Survival." *Proceedings of the American Society of Clinical Oncology* 24: 56.

Kempin, S., C. Cirrincione, D. S. Straus, T. S. Gee, Z. Arlin, B. Koziner, C. Pinsky, L. Nisce, J. Myers, B. J. Lee III, B. D. Clarkson, L. J. Old, and H. F. Oettgen

1981 "Improved Remission Rate and Duration in Nodular Non-Hodgkin's Lymphoma (NNHL) with the Use of Mixed Bacterial Vaccine (MBV)." *Proceedings of the American Society of Clinical Oncology* 22: 514.

Kendall, Arthur Isaac

1931 "Filterable Bodies Seen with the Rife Microscope." *Science* (Supplement: Science News. Anonymous article probably by Kendall.) 74(1928): 10–12.

1932 "The Filtration of Bacteria." *Science* 75(1942): 295–301.

Kendall, Arthur, and Royal Raymond Rife

1931 "Observations on Bacillus Typhosus in its Filterable State." *California and Western Medicine* 35(6): 409–11.

Kendall, Arthur, Hans Zinsser, T. M. Rivers, and William Welch

1932 "Filterable Forms of Bacteria and Their Significance." [Presentation by Kendall with discussion.] *Journal of the American Medical Association* 99(1): 67–69.

Kennaway, Ernest

1955 "The Identification of a Carcinogenic Compound in Coal-Tar." *British Medical Journal* 2: 749–52.

Klieneberger-Nobel, Emmy

1949 "Origin, Development, and Significance of L-Forms in Bacterial Cultures." *Journal of General Microbiology* 3(3): 434–42.

1951 "Filterable Forms of Bacteria." *Bacteriological Review* 15: 77–103.

1960 "L-Forms of Bacteria." In I. C. Gunsalus and Roger Stanier (eds.), *The Bacteria: A Treatise on Structure and Function. Volume I: Structure.* New York: Academic Press.

Kloppenburg, Margreet, Ferdinand Breedveld, Jack Terwiel, Constant Mallee, and Ben Dijkmans

1995 "Minocycline in Active Rheumatoid Arthritis." *Arthritis and Rheumatism* 37(5): 629–35.

Kohler, Robert

1982 *From Medical Chemistry to Biochemistry.* Cambridge: Cambridge University Press.

1985 "Bacterial Physiology: The Medical Context." *Bulletin of the History of Medicine* 59: 54–74.

1991 *Partners in Science: Foundations and Natural Scientists, 1900–1945.* Chicago: University of Chicago Press.

Kosty, Michael, James Herndon, Mark Green, and O. Ross McIntyre

1995 "Placebo-Controlled Randomized Study of Hydrazine Sulfate in Lung Cancer." *Journal of Clinical Oncology* 13: 1529–30.

Krichevsky, Alexander, Elizabeth Campbell-Acevedo, Jennifer Tong, and Hernan Acevedo

1995 "Immunological Detection of Membrane-Associated Human Luteinizing Hormone Correlates with Gene Expression in Cultured Human Cancer and Fetal Cells." *Endocrinology* 136: 1034–39.

Kuhn, Thomas

1970 *The Structure of Scientific Revolutions.* 2nd edition. Chicago: University of Chicago Press.

1977 *The Essential Tension.* Chicago: University of Chicago Press.

Lakatos, Imre

1978 *The Methodology of Scientific Research Programmes.* Cambridge: Cambridge University Press.

Lakhovsky, George

1988 *The Secret of Life: Electricity, Radiation, and Your Body.* Cosa Mesa, Calif.: Noontide Press.

Laquer, G. L., E. G. McDaniel, and H. Matsumoto

1967 "Tumor Induction in Germ-free Rats with Methylazoxymethanol (MAM) and Synthetic MAM Acetate." *Journal of the National Cancer Institute* 39: 355–71.

Latour, Bruno

1987 *Science and Action.* Cambridge: Harvard University Press.

1988 *The Pasteurization of France.* Cambridge: Harvard University Press.

Laudan, Larry

1977 *Progress and Its Problems.* Berkeley: University of California Press.

Lederberg, Joshua, and Edward Tatum

1946 "Gene Recombination in *Escherichia coli.*" *Nature* 158: 558.

Lerner, Michael

1994 *Choices in Healing.* Cambridge: MIT Press.

L'Esperance, Elise

1931 "Studies in Hodgkin's Disease." *Annals of Surgery* 93: 162–68.

Levinson, Warren, J. Michael Bishop, Nancy Quintrell, and Jean Jackson

1970 "Presence of DNA in Rous Sarcoma Virus." *Nature* 227: 1023–25.

Livingston, Afton Monk, Virginia Livingston, Eleanor Alexander-Jackson, and Gerhard Wolter

1970 "Toxic Fractions Obtained from Tumor Isolates and Related Clinical Implications." (Published under Virginia Wuerthele-Caspé.) *Annals of the New York Academy of Sciences* 174(2): 675–89.

Livingston, Virginia

1949 "Mycobacterial Forms Observed in Tumors." (Published under Virginia Wuerthele-Caspé.) *Journal of the American Medical Women's Association* 4: 135–41.

1955 "Neoplastic Infections of Man and Animals." (Published under Virginia Wuerthele-Caspé.) *Journal of the American Medical Women's Association* 10: 261–66.

1972 *Cancer: A New Breakthrough.* (Published under Virginia Wuerthele-Caspé Livingston.) Los Angeles: Nash Publishing.

1979 "The Role of Nutrition in the Immunotherapy of Cancer." *Journal of the International Academy of Preventive Medicine* 5(2): 54–75.

1984 *The Conquest of Cancer: Vaccines and Diet.* (Published under Virginia Livingston-Wheeler, with Edmond Addeo.) New York: Franklin Watts.

1989 "Vaccines for Cancer." Presentation before the annual meeting of the Cancer Control Society, Pasadena, Calif.

Livingston, Virginia, and Eleanor Alexander-Jackson

1965a "An Experimental Biologic Approach to the Treatment of Neoplastic Disease." (Published under Virginia Wuerthele-Caspé Livingston.) *Journal of the American Women's Medical Association* 20(9): 858–66.

1965b "Mycobacterial Forms in Myocardial Vascular Disease." (Published under Virginia Wuerthele-Caspé Livingston.) *Journal of the American Medical Women's Association* 20: 499–552.

1970 "A Specific Type of Organism Cultured from Malignancy: Bacteriology and Proposed Classification." (Published under Virginia Wuerthele-Caspé Livingston.) *Annals of the New York Academy of Sciences* 174: 636–54.

Livingston, Virginia, E. Alexander-Jackson, J. A. Anderson, J. Hillier, R. M. Allen, and L. W. Smith

1950 "Cultural Properties and Pathogenicity of Certain Microorganisms Observed from Various Proliferative and Neoplastic Diseases." (Published under Virginia Wuerthele-Caspé.) *American Journal of the Medical Sciences* 220: 636–48.

Livingston, Virginia, E. Alexander-Jackson, M. Gregory, L. W. Smith, I. C. Diller, and Z. Mankowski

1956 "Intracellular Acid-Fast Microorganisms Isolated from Two Cases of Hepatolenticular Degeneration." (Published under Virginia Wuerthele-Caspé.) *Journal of the American Medical Women's Association* 11(4): 120–29.

Livingston, Virginia, E. Alexander-Jackson, and L. W. Smith

1953 "Some Aspects of the Microbiology of Cancer." (Published under Virginia Wuerthele-Caspé.) *Journal of the American Medical Women's Association* 8(1): 7–12.

Livingston, Virginia, and R. M. Allen

1948 "Presence of Consistently Recurring Invasive Mycobacterial Forms in Tumor Cells." (Published under Virginia Wuerthele-Caspé.) *New York Microscopial Society Bulletin* 2: 5–18.

Livingston, Virginia, E. Brodkin, and C. Mermod

1947 "Etiology of Scleroderma, a Preliminary Report." (Published under Virginia Wuerthele-Caspé.) *Journal of the Medical Society of New Jersey* 44(7): 256–59.

Livingston, Virginia, and Afton Monk Livingston

1972 "Demonstration of *Progenitor Cryptocides* in the Blood of Patients with Collagen and Neoplastic Diseases." *Transactions of the New York Academy of Sciences* Series II 34(5): 433–53.

1974 "Some Cultural, Immunological, and Biochemical Properties of *Progenitor Cryptocides*." (Published as Virginia Wuerthele-Caspé Livingston.) *Transactions of the New York Academy of Sciences.* Series II. 36(6): 569–82.

Livingston, Virginia, and John Majnarich

1986 "Inhibition of Growth of Mouse Sarcoma 180 by Vitamin A and Progenitor Cryptocides Antigens." (Published under Virginia Livingston-Wheeler.) *Journal of Nutrition, Growth, and Cancer* 3: 91–93.

Longino, Helen

1994 "In Search of Feminist Epistemologies." *Monist* 77(4): 472–85.

Löwy, Ilana

1993 "Innovation and Legitimation Strategies: The Story of the New York Cancer Research Institute." In Ilana Löwy (ed.), *Medicine and Change: Historical and Sociological Studies of Medical Innovation.* London: John Libbey and Co.

Lynes, Barry

1987 *The Cancer Cure That Worked.* Queensville, Ont.: Marcus Books.

McGinnis, Lamar

1991 "Alternative Therapies, 1990. An Overview." *Cancer* 67 (6 Supp.): 1788–92.

McGregor, Deborah Kuhn

1989 *Sexual Surgery and the Origins of Gynecology.* New York: Garland Publishing.

McKay, K. A., D. H. Neil, and A. H. Corner

1967 "The Demonstration of a Single Species of an Unclassified Bacterium in Five Cases of Bovine Lymphoma." *Growth* 31: 357–68.

McKendry, Robert

1995 "Is Rheumatoid Arthritis Caused by an Infection?" *Lancet* 345: 1319–20.

MacKenzie, Donald

1981 "Interests, Positivism, and History." *Social Studies of Science* 11: 498–501.

1983 *Statistics in Britain.* Edinburgh: University of Edinburgh Press.

1984 "Reply to Yearly." *Studies in the History and Philosophy* 15(3): 251–59.

McManus, Linda, Michael Naughton, and Antonio Martinez-Hernandez
1976 "Human Chorionic Gonadotropin in Human Neoplastic Cells." *Cancer Research* 36: 3476–81.

Macomber, P. B.
1990 "Cancer and Cell Wall Deficient Bacteria." *Medical Hypotheses* 32: 1–9.

Manning, Kenneth
1983 *Black Apollo of Science.* Oxford: Oxford University Press.

Markle, Gerald, and James Peterson
1980 *Politics, Science, and Cancer: The Laetrile Phenomenon.* Washington, D.C.: AAAS.
1987 "Resolution of the Laetrile Controversy: Past Attempts and Future Prospects." In H. Tristram Engelhardt, Jr., and Arthur L. Caplan (eds.), *Scientific Controversies.* Cambridge: Cambridge University Press.

Marshall, Barry
1994 *"Helicobacter Pylori." American Journal of Gastroenterology* 89: S116–S128.

Marshall, Barry, J. Robin Warren, Elizabeth Blincow, Michael Phillips, C. Stewart Goodwin, Raymond Murray, Stephen Blackbourn, Thomas Waters, and Christopher Sanderson
1988 "Prospective Double-Blind Trial of Duodenal Ulcer Relapse after Eradication of *Campylobacter Pylori.*" *Lancet* 8626: 1437–41.

Marshino, O.
1944 "Administration of the National Cancer Institute Act, August 5, 1937 to June 30, 1943." *Journal of the National Cancer Institute* 4: 429–43.

Martin, Brian
1993 "The Critique of Science Becomes Academic." *Science, Technology, and Human Values* 18(2): 247–59.
1996 "Sticking a Needle into Science: The Case of Polio Vaccines and the Origin of AIDS." *Social Studies of Science* 26: 245–76.

Martin, Brian, C. M. Ann Baker, Clyde Manwell, and Cedric Pugh (eds.)
1986 *Intellectual Suppression.* London: Angus and Robertson.

Martin, Emily
1987 *The Woman in the Body.* Boston: Beacon Press.
1991 "The Egg and the Sperm: How Science has Constructed a Romance Based on Stereotypical Male-Female Roles." *Signs* 16(3): 485–501.
1994 *Flexible Bodies.* Boston: Beacon Press.

Martin, Hayes
1944 "Interview with Mr. George Barclay—Memorial Hospital." Rockefeller Archives, Hayes-Martin Collection, RG 500, Box 6, Folder 106.

Maruo, Takeshi, Herman Cohen, Sheldon Segal, and S. S. Koide
1979 "Production of Choriogonadotropin-like Factor by a Microorganism." *Proceedings of the National Academy of Science* 76: 6622–26.

Mattman, Lida

1974 *Cell Wall Deficient Forms.* Cleveland: CRC Press.

1993 *Cell Wall Deficient Forms.* 2nd edition. Boca Raton: CRC Press.

Mazet, Georges

1941 "Étude bactériologique sur le maladie d'Hodgkin." *Extrait de Montpellier Médicale,* juillet-aoòt.

Mellon, Ralph, and L. W. Fisher

1932 "New Studies on the Filterability of Pure Cultures of the Tubercle Group of Microorganisms." *Journal of Infectious Disease* 51: 117–28.

Miller, T. N., and J. T. Nicholson

1971 "End Results in Reticulum Cell Sarcoma of Bone Treated by Toxin Therapy Alone or Combined with Surgery and/or Radiation (47 Cases) or with Concurrent Infection (5 cases)." *Cancer* 27: 514–48.

Miner, Roy Waldo (ed.)

1952 "Viruses as Causative Agents in Cancer." *Annals of the New York Academy of Sciences* 54(6): 869–1232.

Moss, Ralph

1989 *The Cancer Industry.* New York: Paragon.

1992 *Cancer Therapy: The Independent Consumer's Guide to Non-Toxic Treatment and Prevention.* New York: Equinox Press.

1995 *Questioning Chemotherapy.* New York: Equinox Press.

1996 *The Cancer Industry.* 2nd edition. New York: Equinox Press.

Muir, C. S., J. F. Fraumeni, and R. Doll

1994 "The Interpretation of Time Trends." *Cancer Surveys* 20: 5–21.

Mullins, Nicholas

1972 "The Development of a Scientific Specialty: The Phage Group and the Origins of Molecular Biology." *Minerva* 10 (Jan.): 52–82.

National Cancer Institute

1991 *Cancer Statistics Review 1973–1988.* NIH Publication No. 91–2789. Bethesda, Md.: National Institutes of Health.

Naughton, Michael, Deborah Merrill, Linda McManus, Louis Fink, Edward Berman, Miriam Jill White, and Antonio Martinez-Hernandez

1975 "Localization of the Beta Chain of Human Chorionic Gonadotropin on Human Tumor Cells and Placental Cells." *Cancer Research* 35: 1887–90.

Nauts, Helen Coley

1975 *Osteogenic Sarcoma: End Results Following Immunotherapy with Bacterial Vaccines, 165 Cases or Following Bacterial Infections, Inflammation, or Fever, 41 Cases.* Monograph #15. New York: Cancer Research Institute.

1976 "Immunotherapy of cancer by Bacterial Vaccines." Paper presented at the International Symposium on Detection and Prevention of Cancer. New York, April 25–May 1.

1980 *The Beneficial Effects of Bacterial Infections on Host Resistance to Cancer. End Results in 449 Cases.* Monograph No. 8, 2nd edition. New York: Cancer Research Institute.

1995 *Coley Toxins: 1893–1995 and Beyond*. New York: Institute for Cancer Research.

Netterberg, Robert, and Robert Taylor

1981 *The Cancer Conspiracy.* New York: Pinnacle Books.

Nonclerq, Marie

1982 *Antoine Béchamp, 1816–1908.* Paris: Maloine.

Nuzum, John

1921 "A Critical Study of an Organism Associated with a Transplantable Carcinoma of the White Mouse." *Surgery, Gynecology, and Obstetrics* 33: 167–76.

1925 "The Experimental Production of Metastasizing Carcinoma in the Breast of the Dog and Primary Epithelioma in Man by Repeated Inoculation of a Micrococcus Isolated from Human Breast Cancer." *Surgery, Gynecology, and Obstetrics* 40: 343–52.

Oettgen, Herbert, and Lloyd Old

1991 "The History of Cancer Immunotherapy." In Vincent DeVita Jr., Samuel Hellman, and Steven Rosenberg (eds.), *The Biologic Therapy of Cancer.* New York: J. B. Lippincott.

Office of Technology Assessment of the U.S. Congress

1990 *Unconventional Cancer Treatments,* OTA-H-405. Washington, D.C.: U.S. Government Printing Office.

Old, Lloyd J.

1988 "Tumor Necrosis Factor." *Scientific American* May: 59–75.

Parsonnet, Julie, Svein Hansen, Larissa Rodriguez, Arnold Gelb, Roger Warnke, Egil Jellum, Norman Orentreich, Joseph Vogelman, and Gary Friedman

1994 "*Helicobacter Pylori* Infection and Gastric Lymphoma." *New England Journal of Medicine* 330: 1267–71.

Patterson, James

1987 *The Dread Disease: Cancer and Modern American Culture.* Cambridge: Harvard University Press.

Pauling, Linus

1993 *Cancer and Vitamin C.* Philadelphia: Camino Books.

Paulus, Harold

1995 "Reply." *Annals of Internal Medicine* 123(5): 393.

Payer, Lynn

1988 *Medicine and Culture: Varieties of Treatment in the United States, England, West Germany, and France.* New York: Henry Holt and Co.

Pease, Phyllis

1970 "Discussion: Microorganisms Associated with Malignancy." *Annals of the New York Academy of Sciences* Vol. 174, Art. 2: 782–85.

Pelton, Ross, and Lee Overholser

1994 *Alternatives in Cancer Therapy.* New York: Simon and Schuster.

Peterson, James, and Gerald Markle

1979a "The Laetrile Controversy." In Dorothy Nelkin (ed.), *Controversy: Politics of Technical Decisions.* Beverly Hills: Sage.

1979b "Politics and Science in the Laetrile Controversy." *Social Studies of Science* 9: 139–66.

Popper, Karl

1963 *Conjectures and Refutations.* London: Routledge.

Price, J. Townley, and Glenn Bulmer

1972 "Tumor Induction by *Cryptococcus Neoformans.*" *Infection and Immunity* 6(2): 199–205.

Priebe, Waldemar (ed.)

1995 *Anthracycline Antibiotics.* Washington, D.C.: American Chemical Society.

Proctor, Robert

1995 *Cancer Wars.* Cambridge: Harvard University Press.

Rappin, Gustave

1939 *Observations sur les granulations colloïdales de la cellule cancéreuse.* Nantes: Imprimerie de Bretagne.

Räth, C.

1925 "Über das Vorkommen im Mikro-organismen in Tumoren." *Zeitschrift für Angewandte Chemie* July 23.

Regelson, William

1995 "Have We Found the 'Definitive Cancer Biomarker'?" *Cancer* 76: 1299–1301.

Rettger, Leo, and Hazel Gillespie

1933 "Bacterial Variation, with Special Reference to Pleomorphism and Filtrability." *American Journal of Bacteriology* 26: 289–318.

Rettig, Richard

1977 *Cancer Crusade: The Story of the National Cancer Act of 1971.* Princeton: Princeton University Press.

Richards, Evelleen

1981 *Vitamin C and Cancer.* New York: St. Martin's Press.

Richier-Chevrel, M.-E.

1951 "Recherche sur les bactériémies chez les cancéreux: contribution à l'étude du charlatanisme médical et de l'insuffisance des lois françaises pour le combattre," thesis defended before the Faculté de Paris and supervised by Dr. Prévost, chef de service de l'Institut Pasteur.

Ries, L. A. G., B. F. Hankey, and B. K. Edwards

1990 *Cancer Statistics Review 1973–1987*. NIH Publication 90–2789. Bethesda, Md.: National Institutes of Health.

Rife, Royal Raymond, and John Crane

1953 "History of the Development of a Successful Treatment for Cancer and Other Virus, Bacteria, and Fungi." Unpublished manuscript, U.S. National Library of Medicine.

Riordan, M. H., H. D. Riordan, X. Meng, Y. Li, and J. A. Jackson

1994 "Intravenous Ascorbate as a Tumor Cytotoxic Chemotherapeutic Agent." *Medical Hypotheses* 44: 207–13.

Rook, Graham

1992 "Tumors and Coley's Toxins." *Nature* 357: 545.

Rosenberg, Steven, and John Barry

1992 *The Transformed Cell: Unlocking the Mysteries of Cancer.* New York: Avon.

Rosenow, Edward C.

1914 "Transmutations within the Streptococcus-Pneumococcus Group." *Journal of Infectious Disease* 14(1): 1–32.

1932 "Observations with the Rife Microscope of Filter-Passing Forms of Microorganisms." *Science* 76(1965): 192–93.

Rossiter, Margaret W.

1993 "The Matthew Matilda Effect in Science." *Social Studies of Science* 23: 325–41.

Rous, Peyton

1910 "A Transmissible Avian Neoplasm (Sarcoma of the Common Fowl)." *Journal of Experimental Medicine* 12: 696–705.

1941 "Virus Relationships to Tumors." In Louis Fieser, S. P. Reimann, Peyton Rous, W. H. Lewis, Margaret Lewis, and Baldwin Lake (eds.), *Cause and Growth of Cancer: University of Pennsylvania Bicentennial Conference.* Philadelphia: University of Pennsylvania.

Rusch, Harold

1985 "The Beginnings of Cancer Research in the United States." *Journal of the National Cancer Institute* 74(2): 391–403.

Sahlins, Marshall

1976 *Culture and Practical Reason*. Chicago: University of Chicago Press.

Sale, David

1995 *Overview of Legislative Developments Concerning Alternative Health Care in the United States.* Kalamazoo, Mich.: Fetzer Institute.

Scammell, Henry

1993 *The Arthritis Breakthrough*. New York: M. Evans and Company.

Sclove, Richard

1995 *Democracy and Technology.* New York: Guilford Press.

1996 "Democratizing Science Advisory Panels?" Loka Alert 3: 3, http://www.amherst.edu/loka. Amherst, Mass.: Loka Institute.

Scott, Michael

1926 "Clinical Experiences with Carcinoma Antitoxin." *Irish Journal of Cancer* 3(9): 1–6.

Scott, Pam, Evelleen Richards, and Brian Martin

1990 "Captives of Controversy: The Myth of the Neutral Social Researcher in Contemporary Scientific Controversies." *Science, Technology, and Human Values* 15 (4): 474–94.

Seibert, Florence

1968 *Pebbles on the Hill of a Scientist.* St. Petersburg, Fla.: Author.

Seibert, Florence, J. A. Baker, J. Kierking, R. Abadal, and R. L. Davis

1973 "Decrease in Spontaneous Tumors by Vaccinating C3H Mice with an Homologous Bacterial Vaccine." *International Research Communication Systems* March, p. 53.

Seibert, Florence, and Robert Davis

1977 "Delay in Tumor Development Induced with a Bacterial Vaccine." *Journal of the Reticuloendothelial Society* 21(4): 279–82.

Seibert, Florence B., F. M. Feldman, R. L. Davis, and I. S. Richmond

1970 "Morphological, Biological, and Immunological Studies on Isolates from Tumors and Leukemic Bloods." *New York Academy of Science* 174(2): 690–728.

Seidel, R. E., and M. Elizabeth Winter

1944 "The New Microscopes." *Journal of the Franklin Institute* 237(2): 103–30.

Sharma, Ursula

1992 *Complementary Medicine Today: Practitioners and Patients.* London and New York: Routledge.

Shear, M. J., Floyd Turner, Adrien Perrault, and Theresa Shovelton

1943 "Chemical Treatment of Tumors. V. Isolation of the Hemorrhage-producing Fraction from *Serratia marcescens (B. prodigiosum)* Culture Filtrate." *Journal of the National Cancer Institute* 4: 81–97.

Shimkin, Michael

1977 "As Memory Serves—An Informal History of the National Cancer Institute, 1937–57." *Journal of the National Cancer Institute* 59(2): 559–600.

Shwartzman, G., and N. Michailovsky

1932 "Phenomenon of Local Tissue Reactivity to Bacterial Filtrates in Treatment of Mouse Sarcoma 180." *Proceedings of the Society for Experimental Biology and Medicine* 29: 737–41.

Star, Susan Leigh
1995 *Ecologies of Knowledge.* Albany: State University of New York Press.

Starnes, Charlie
1992 "Coley's Toxins in Perspective." *Nature* 357: 11–12.

Starr, Paul
1982 *The Social Transformation of American Medicine.* New York: Basic Books.

Stoddard, George
1955 *"Krebiozen": The Great Cancer Mystery.* Boston: Beacon Press.

Stokes, Ray
1988 *Divide and Prosper: The Heirs of I. G. Farben under Allied Authority, 1945–51.* Berkeley and Los Angeles: University of California Press.

Strickland, Stephen
1972 *Politics, Science, and Dread Disease.* Cambridge: Harvard University Press.

Studer, Kenneth, and Daryl Chubin
1980 *The Cancer Mission: Social Contexts of Biomedical Research.* Beverly Hills: Sage.

Tang, Z. Y., H. Y. Zhou, G. Zhao, L. M. Chai, M. Zhou, J. Z. Lu, K. D. Liu, H. F. Havas, and H. C. Nauts
1991 "Preliminary Result of Mixed Bacterial Vaccine as Adjuvant Treatment of Hepatocellular Carcinoma." *Medical Oncology and Tumor Pharmacology* 8: 23–28.

Thomas, Gordon
1975 *Issels: The Biography of a Doctor.* London: Hodder and Stoughton. (American edition: *Dr. Issels and His Revolutionary Cancer Treatment,* New York: Peter Wyden, 1973.)

Tilley, Barbara, Graciela Alarcón, Stephen Heyse, et al.
1995 "Minocycline in Rheumatoid Arthritis." *Annals of Internal Medicine* 122(2): 81–89.

Torrey, J. C.
1916 "Bacteria Associated with Certain Types of Abnormal Lymph Glands." *Journal of Medical Research* 34: 65–80.

Traweek, Sharon
1992 "Border Crossings: Narrative Strategies in Science Studies and among Physicists in Tsukuba Science City, Japan." In Andrew Pickering (ed.), *Science as Practice and Culture.* Chicago: University of Chicago Press.

Treichler, Paula
1991 "How to Have Theory in an Epidemic: The Evolution of AIDS Treatment Activism." In Constance Penley and Andrew Ross (eds.), *Technoculture: Cultural Politics, Vol. 3.* Minneapolis: University of Minnesota Press.

Triozzi, Pierre, Diane Gochnour, Edward Martin, Wayne Aldrich, John Powell, Julian Kim, Donn Young, and John Lombardi

1994 "Clinical and Immunologic Effects of a Synthetic Beta-Human Chorionic Gonadotropin Vaccine." *International Journal of Oncology* 5: 1447–53.

Varmus, Harold, and Robert Weinberg

1993 *Genes and the Biology of Cancer.* New York: Scientific American Library.

Villequez, Ernest

1955 *Le parasitisme latent des cellules du sang chez l'homme, en particulier dans le sang des cancereux.* Paris: Librarie Maloine.

1969 *Le cancer de l'homme: L'étude interdite.* Paris: Delta.

Wade, Nicholas

1971 "Special Virus Program: Travails of a Biological Moonshot." *Science* 174(Dec. 24): 1306–11.

Walker, Martin

1993 *Dirty Medicine: Science, Big Business, and the Assault on Natural Health Care.* London: Slingshot Publications.

Wallis, Roy

1985 "Science and Pseudo-Science." *Social Science Information* 24(3): 585–601.

Walters, Richard

1993 *Options: The Alternative Cancer Therapy Book.* Garden City Park, N.Y.: Avery Publishing Group.

Ward, Patricia Spain

1984 "Who Will Bell the Cat?" *Bulletin of the History of Medicine* 58: 28–52.

1996 "History of BCG Vaccine." *Townsend Letter for Doctors and Patients* October, pp. 72–77.

Weber, Max

1958 *The Protestant Ethic and the Spirit of Capitalism.* New York: Charles Scribner's Sons.

Weinberg, Robert

1994 "Oncogenes and Tumor Suppressor Genes." *CA—A Cancer Journal for Clinicians* 44(3): 160–70.

Weinman, David, Edward Johnston, Chairoj Saeng-udom, J. A. Whitaker, Poonsri Tamasatit, Kampol Panas-Ampol, and Eleanor Fort

1968 "Lymphoma: Intranuclear Bacilliform Structures in a Patient with Febrile Anemia." *American Journal of Pathology* 52(6): 1129–43.

White, John

1953 "Report on One Hundred Proven Cases of Malignancy Treated by a Specific Antiserum." *Atti del VI Congresso Internazionale di Microbiologia* Vol. 6, sec. 17A, pp. 29–40.

White, Milton

1965 "Etiology of Malignancies." *Journal of the International College of Surgeons* Section I. 54(6): 593–602.

Wiemann, Bernadette, and Charlie Starnes

1994 "Coley's Toxins, Tumor Necrosis Factor, and Cancer Research: A Historical Perspective." *Pharmacology and Therapeutics* 64: 529–64.

Wing, Phyllis, Tony Tong, and Sherry Bolden

1995 "Cancer Statistics, 1995." *CA—A Cancer Journal for Clinicians* 45(1): 8–30.

Wolbach, Simeon Burt

1947 "Hans Zinsser." In *National Academy of Sciences of the United States of America: Biographical Memoirs.* Vol. 24. Washington, D.C.: National Academy of Sciences.

Woolgar, Steve

1981a "Critique and Criticism: Two Readings of Ethnomethodology." *Social Studies of Science* 11: 504–14.

1981b "Interests and Explanation in the Social Study of Science." *Social Studies of Science* 11: 365–94.

Wotherspoon, Andrew, Claudio Doglioni, Michele de Boni, Jo Spencer, and Peter Isaacson

1994 "Antibiotic Treatment for Low-Grade Gastric MALT Lymphoma." *Lancet* 343(June 11): 1503.

Wright, Karen

1991 "Going by the Numbers." *New York Times Sunday Magazine* December 15, pp. 58–79.

Wuerthele-Caspé, Virginia. (*See* Livingston, Virginia)

Wyburn-Mason, Roger

1964 *A New Protozoan: Its Relation to Malignant and Other Diseases.* London: Charles C. Thomas.

Yamagiwa, Katsusaburo, and Koichi Ichikawa

1916 "Experimentelle Studie über die Pathogenese der Epithelialgeschwülste." *Mitteilungen aus der medizinischen Fakultät der kaiserlichen Universität zu Tokyo* 15: 296–344.

Yaremchuk, William

1977 *The Cancer War: The Movement to Establish the National Cancer Institute, 1927–1937.* Author.

Yearley, Stephen

1982 "The Relationship Between Epistemological and Sociological Cognitive Interests: Some Ambiguities Underlying the Use of Interest Theory in the Study of Scientific Knowledge." *Studies in the History and Philosophy* 13(4): 353–88.

Young, James

1925a "A New Outlook on Cancer, Irritation, and Infection." *British Medical Journal* Jan. 10, pp. 60–64.

1925b "Letter to the Editor." *British Medical Journal* Aug. 8, p. 271.

Zheren, Guo, and Helen Coley Nauts

1991 "Pilot Study of Mixed Bacterial Vaccine (MBV) and Pediatric Cancers: 52 Cases." New York: Institute for Cancer Research.

Zuckerman, Harriet, and Joshua Lederberg

1986 "Postmature Scientific Discovery?" *Nature* 324(Dec. 18/25): 629–31.

Index

About the Author

David J. Hess is an anthropologist and tenured professor in the Science and Technology Studies Department at Rensselaer Polytechnic Institute. His previous research on alternative medicine involved ethnographic fieldwork with the Spiritist movement of Brazil, where he examined beliefs, practices, and scientific controversies associated with spirits, mediums, and healing. That research led to *Samba in the Night, Spirits and Scientists, The Brazilian Puzzle* (a book coedited with Roberto DaMatta), and *Science in the New Age,* a comparative book on similar issues in the United States. Hess has also written two general books on science studies: *Science and Technology in a Multicultural World* and *Science Studies: An Advanced Introduction.* He is the recipient of various grants and awards, including two Fulbrights and a National Science Foundation grant for research on the public understanding of science, and he is chair of the Committee of the Anthropology of Science, Technology, and Computing of the American Anthropological Association. *Can Bacteria Cause Cancer?* is the first in a planned series of books on alternative cancer therapies, cultural and political issues related to their controversial status, and the problem of how to evaluate them.